Artificial Neural Networks for System Identification and Control

In an age where intelligent systems are transforming engineering practice, *Artificial Neural Networks for System Identification and Control* offers a clear roadmap to mastering artificial intelligence (AI)-driven modeling and control. From mathematical neuron models to adaptive Artificial Neural Network (ANN)-based controllers, this book combines theory, algorithms, and hands-on coding to help readers design and analyze intelligent systems. Rich with visual examples and real-world case studies, it demonstrates how neural networks outperform traditional control methods in handling nonlinearity, uncertainty, and dynamic system behavior.

- Offers a practical and accessible guide to ANN-based system identification and control
- Blends mathematical insight with real engineering applications
- Provides Python-supported examples and visual case studies
- Highlights key advances in nonlinear modeling and adaptive control design
- Bridges the gap between theory, simulation, and real-world deployment

This book is intended for engineers, researchers, and advanced students seeking to apply artificial intelligence to control theory, robotics, and signal processing and to design smarter, more adaptive engineering systems.

Artificial Neural Networks for System Identification and Control

Serhat Seker, Tahir Cetin Akinci and
Alfredo A. Martinez-Morales

CRC Press
Taylor & Francis Group
Boca Raton London New York

CRC Press is an imprint of the
Taylor & Francis Group, an **informa** business

Designed cover image: iStock

First edition published 2026
by CRC Press
2385 NW Executive Center Drive, Suite 320, Boca Raton FL 33431

and by CRC Press
4 Park Square, Milton Park, Abingdon, Oxon, OX14 4RN

CRC Press is an imprint of Taylor & Francis Group, LLC

ISBN: 978-1-041-04911-1 (hbk)
ISBN: 978-1-041-14338-3 (pbk)
ISBN: 978-1-003-67397-2 (ebk)

DOI: 10.1201/9781003673972

Typeset in Times
by codeMantra

Contents

Preface

Artificial Neural Networks (ANNs) have transformed the landscape of system identification and control, offering unparalleled capabilities in handling nonlinear systems, optimizing performance, and adapting to various operational conditions. This book, *Artificial Neural Networks for System Identification and Control*, aims to provide a comprehensive understanding of the theoretical foundations, mathematical models, and practical applications of ANNs in these critical areas.

The inspiration for this book stems from the need to bridge the gap between complex neural network theories and their practical implementations in system identification and control. As researchers and practitioners across various fields of engineering, we have observed a growing interest and significant advancements in this domain. Our objective is to equip readers with the knowledge and tools necessary to harness the power of ANNs for real-world applications.

Throughout this book, we delve into the intricate workings of artificial neurons, explore several neural network topologies, and elucidate the underlying mathematical models. We also present a detailed examination of algorithms, such as backpropagation, which is used in both feedforward and recurrent neural network topologies, highlighting their relevance in system identification and control tasks. By integrating theoretical concepts with practical examples and case studies, we aim to make this book a valuable resource for students, researchers, and professionals alike.

We would like to extend our gratitude to our colleagues and collaborators who have contributed to the development of this book. Their insights and feedback have been instrumental in shaping the content and ensuring its relevance to contemporary challenges in the field. We also appreciate the support of our respective institutions, Istanbul Technical University (ITU) and the University of California Riverside (UCR), for providing an environment conducive to academic research and innovation.

We hope that this book will inspire further research and advancements in the application of ANNs, fostering a deeper understanding and appreciation of their potential in system identification and control.

Serhat Seker
Tahir Cetin Akinci
Alfredo A. Martinez-Morales

Authors

Serhat Seker currently serves as a Full Professor in the Department of Electrical Engineering at Istanbul Technical University (ITU). He completed his undergraduate studies in electrical engineering and pursued his graduate education in electrical and nuclear engineering. His academic and research interests focus on cognitive systems, data processing, signal analysis, and the artificial intelligence–based monitoring of energy systems. Professor Seker has extensive international research experience, having collaborated with research groups at the University of Tennessee in the United States and the Energy Research Centre of the Netherlands. In recognition of his scientific contributions, he was awarded the title of Honorary Professor by the National University of Radio Electronics (NURE) in Ukraine. Between 2014 and 2020, he served as the Dean of the Faculty of Electrical and Electronics Engineering at ITU, where he played a significant role in academic and administrative development. He currently teaches in the Department of Electrical Engineering at ITU and at ESIGELEC Graduate School of Engineering in France. In addition, he provides academic consultancy for the establishment and development of the Electrical Engineering Department at the University of Djibouti in Africa, where he also teaches. Alongside his scientific research, Professor Seker is an active author, having published ten books, including one textbook, as well as numerous articles, conference papers, and seminar contributions.

Tahir Cetin Akinci (Senior Member, IEEE) earned a bachelor's degree in electrical engineering in 2000, followed by a master's and a PhD in 2005 and 2010, respectively. From 2003 to 2010, he served as a Research Assistant in the Department of Electrical Engineering at Marmara University in Istanbul, Türkiye. In 2020, he was promoted to Full Professor at Istanbul Technical University (ITU), where he also held administrative appointments, including Associate Dean of the Graduate School and the Faculty of Electrical and Electronics Engineering. Dr. Akinci is currently engaged in academic and research activities at the University of California, Riverside (UCR), within the Bourns College of Engineering Center for Environmental Research and Technology (CE-CERT) and Florida Polytechnic University. His research interests include artificial neural networks, deep learning, machine learning, cognitive systems, signal processing, power systems, renewable energy systems, and data analytics. He continues to contribute to both theoretical and applied research in intelligent systems and data-driven modeling.

Alfredo A. Martinez-Morales serves as the Managing Director of the Southern California Research Initiative for Solar Energy (SC-RISE) and holds the position of Research Professor at the Bourns College of Engineering Center for Environmental Research and Technology (CE-CERT). He earned a BS, an MS, and a PhD in electrical engineering at the University of California, Riverside (UCR) in 2005, 2008, and 2010, respectively. His current research focuses on solar cells, alkali metal ion batteries, highly integrated renewables, energy storage systems, and microgrids. Dr. Martinez-Morales plays a key role as a principal investigator in the Sustainable Integrated Grid Initiative and the Distributed Energy Resources Laboratory at UCR, contributing to the engineering, permitting, and deployment of microgrids throughout Southern California.

1 Introduction

1.1 SYSTEM CONCEPT

In terms of system classification, a Linear Time Invariant (LTI) system approach in discrete time is considered. When an excitation signal is applied to the system's input, and the system's response is measured at the output, it is described with only one input and one output terminal. This type of system is generally referred to as a Single-Input Single-Output (SISO) system. On the other hand, a system with two or more input terminals and/or two or more output terminals is also known as a multi-input-multi-output (MIMO) system. In this context, the relevant figures are depicted in the following figures.

Figure 1.1 illustrates two fundamental system architectures used in signal processing and control applications: SISO and MIMO systems. A SISO system processes a single-input signal to generate a single output, making it suitable for simple control mechanisms and traditional signal processing tasks. In contrast, a MIMO system handles multiple inputs and produces multiple outputs, enabling it to process complex data streams simultaneously. MIMO systems are particularly useful in modern applications such as wireless communication, robotics, and advanced control systems, where multiple variables must be analyzed and optimized together.

A system can be categorized as either a continuous-time system or a discrete-time system, as mentioned above. A system is classified as a discrete-time system when it processes discrete-time signals as input and generates a discrete-time response at its output. In the case of the SISO system depicted in the figure above, $x[n]$ represents the input, $y[n]$ represents the system output, and 'n' indicates the discrete-time index.

For a given discrete-time system in the time domain, its response can be effectively represented using convolution, particularly for LTI systems, as given below:

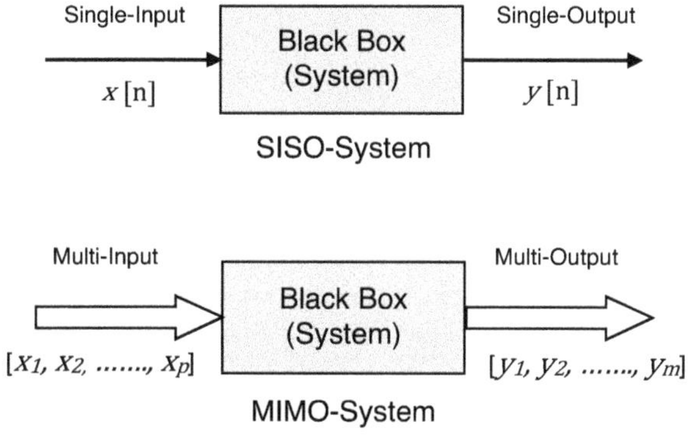

FIGURE 1.1 SISO and MIMO systems in discrete time.

DOI: 10.1201/9781003673972-1

$$y[n] = x[n]*h[n] \tag{1.1}$$

or

$$y[n] = \sum_k x[k]\, h[n-k] \tag{1.2}$$

By applying the Z-Transform to Equation (1.2), it can be represented in the $\mathcal{Z}$ domain. In the frequency domain, the convolution term is transformed into a simple multiplication operator with the assistance of the z:

$$\mathcal{Z}\{y[n]\} = \mathcal{Z}\{x[n]*h[n]\} \tag{1.3}$$

or

$$Y(z) = X(z).H(z) \tag{1.4}$$

Here, $H(z)$ represents the system transfer function in the discrete-time domain. However, neural networks (NNs) can be categorized as SISO, MIMO, and MISO systems, utilizing the non-linear mapping characteristics of the system approach independently of the transfer function concept, as demonstrated in classical systems.

In time domain system studies, the system function is described as the impulse response function, i.e., h[n]. In the NN approach, there is no need for the system function. Additionally, for hardware implementations of conventional systems, mathematical operators such as summation, multiplication, and delay terms may be required for definition. In contrast, NN algorithms are implemented through software and can be easily integrated into embedded systems.

In nature, most systems exhibit non-linear characteristics. However, in engineering systems, non-linear systems are often linearized and analyzed using well-defined analysis techniques for practical solutions, such as state-space representation. Various techniques and direct methods exist for analyzing non-linear systems, including the Lyapunov approach. Linearization can be achieved in several ways, with the simplest and most well-known method being through the Taylor expansion. As a non-numerical example, if we consider a non-linear function $f(x)$, the Taylor expansion of the function $f(x)$ at a point x_0 is given as

$$f(x) = f(x_0) + \frac{f'(x_0)}{1!}(x - x_0) + \frac{f''(x_0)}{2!}(x - x_0)^2 + \cdots \tag{1.5}$$

Consequently, a general linear equation is essentially represented by considering the first two terms of Equation (1.5). In this scenario, if we focus solely on the first two terms of the preceding Taylor expansion and disregard the higher-order terms, it simplifies to

$$f(x) = f(x_0) + \frac{f'(x_0)}{1!}(x - x_0) + \text{Error term} \tag{1.6}$$

or it becomes at around $x_0 = 0$, ignoring the error terms with higher-order terms:

$$y = n + mx \tag{1.7}$$

where $f(x_0)$ is constant n, and also the derivation of the $f(x_0)$ is slope m. Hence, it can be regarded as an approximate linear form of the non-linear function.

Nonetheless, by neglecting the higher-order terms in the expansion, significant error terms may arise. Despite this, the linearization approach is still employed due to its simplicity in computation and analysis. However, the NN approach offers greater flexibility in using non-linear systems in their original, direct form.

1.2 DESCRIPTIONS OF THE NEURAL NETS

The configuration of the MIMO system is shown in Figure 1.2.

Here, $[x_1, x_2, \ldots, x_n]$ and $[y_1, y_2, \ldots, y_m]$ represent the discrete $(n \times 1)$ input and $(m \times 1)$ outputs, respectively. Additionally, a combination of the neurons forms a *Parallel Distributed Structure* as indicated in Figure 1.3.

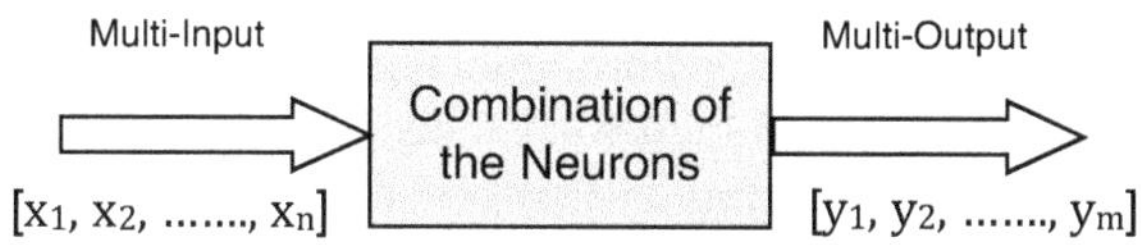

FIGURE 1.2 Block diagram of NN as MIMO systems.

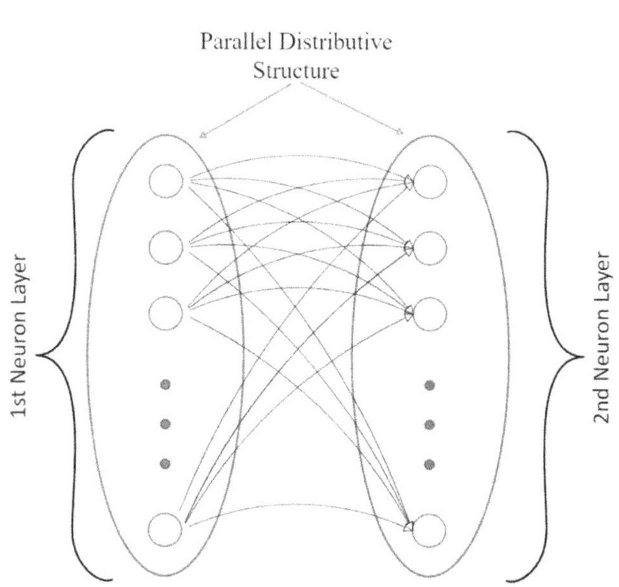

FIGURE 1.3 Parallel distributed connections of the neurons between two layers.

1.3 THE ADVANTAGE OF THE PARALLEL DISTRIBUTIVE STRUCTURE

An NN topology simultaneously possesses several distinctive capabilities, which are closely associated with high reliability, computational speed, and optimization characteristics supported by the learning process. These attributes are elaborated upon as follows:

Reliability: NNs exhibit remarkable reliability. Even in cases where certain connections are disrupted or removed, the network remains functional and operational.

Optimization: The strength of each connection within the network is intricately tied to its weight factor. Optimization is achievable through the implementation of learning algorithms that leverage the sensitivity of these weight factors.

Speed: The parallel and distributed nature of NNs, combined with specialized training algorithms such as the Levenberg–Marquardt Algorithm, enables rapid processing.

Given these considerations, the basic NN configuration, also known as a multi-layer feed forward NN (where information flows from input to output), with '*n*' input nodes and '*m*' output nodes is shown in Figure 1.4. This figure depicts a multi-layer feed-forward NN, a fundamental structure in artificial intelligence and machine learning. The network consists of three main parts: an input layer that receives data, hidden layers where the network learns patterns, and an output layer that produces the final predictions. Each connection between layers represents a transformation of information, allowing the network to recognize complex relationships in the data. The presence of multiple hidden layers enables the model to capture intricate patterns,

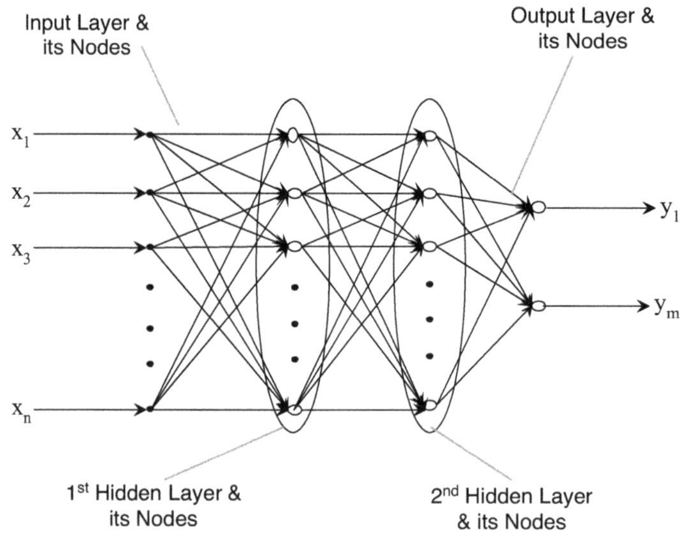

FIGURE 1.4 Multi-layer feed forward NN with two hidden layers.

making it highly effective for applications such as image recognition, speech processing, and predictive analytics.

The most important description of the NN topology is the Determination of Hidden Layers and Hidden Nodes. In this framework, the topological and parametric determinations are done.

1.3.1 Topological Determinations for the Neural Nets

The processing time of an NN is directly influenced by the number of hidden layers and nodes it comprises. A higher count of hidden layers and nodes results in slower network operation due to the increased computational load. Conversely, a network with fewer hidden nodes can achieve faster processing but may not be able to optimize the learning process fully. Therefore, there is a need for topological optimization of the NN structure in addition to parametric optimization.

To simplify the problem, let's consider a scenario with only one hidden layer. In this case, the network's topology is represented in Figure 1.5.

Generally, there is no specific criterion for optimizing the number of hidden nodes. However, it can be determined through trial and error. Nonetheless, one method for determining the optimal number of hidden nodes is by considering Shannon's Information Criteria. Shannon, an American engineer, defined information criteria for communication channel capacity in 1948. Sometimes, Shannon's criteria are used as a unit of measurement for information. The unit of measurement for information is a 'bit.' This approach calculates the amount of information being processed in the hidden layer and can be described in terms of bits. More details on this concept are provided in Section 1.3.

Therefore, the relationship between the concept of bits and the number of hidden nodes is associated with the amount of information processed in the network.

For a given NN structure, there are basically two steps, i.e.,

- Learning
- Test (Recalling)

Thus, optimized topology shows a good performance after the learning and testing steps.

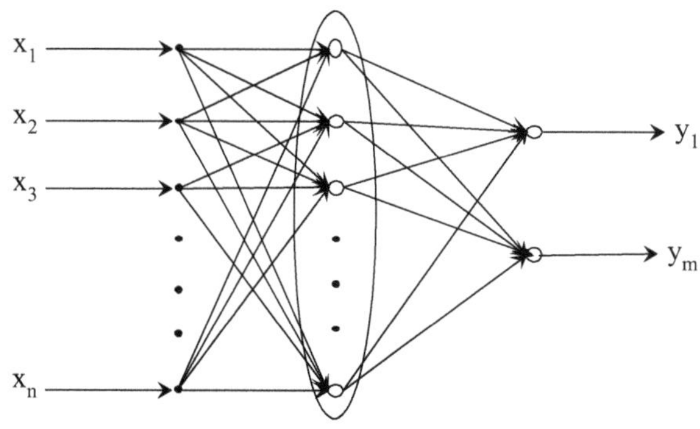

FIGURE 1.5 Multi-layer feed forward NN with a single hidden layer.

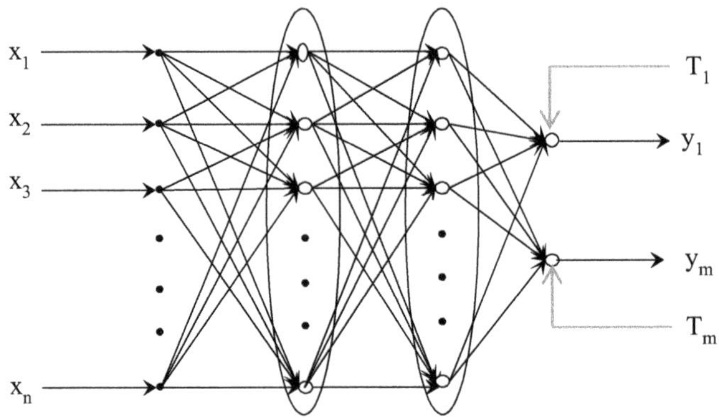

FIGURE 1.6 Multi-layer feed forward NN with a target value (desired output).

1.3.2 Parametric Determinations

This determination is related to the learning parameters and weight factors, which are proportional to the connection strengths of the Multi-Layered Feed Forward NN.

At first, the learning procedure with target values (desired output) is considered. In this context, a general NN has '*n*' input nodes, '*m*' output nodes, and '*m*' target values as shown in Figure 1.6.

In terms of vectors, all variables and vectors can be defined as follows:

Vector $\tilde{x} = [x_1, x_2, \ldots, x_n]$: Input Vector
Vector $\tilde{y} = [y_1, y_2, \ldots, y_m]$: Output Vector
Vector $\tilde{T} = [T_1, T_2, \ldots, T_m]$: Target Value (Desired o/p)
Matrix $\tilde{W} = [W_{ji}]$
$= \tilde{W}_1 \Rightarrow nxk$: Size of i/p and hidden layer
$= \tilde{W}_1 \Rightarrow kxm$: Size of hidden layer and o/p
(where k is the number of hidden nodes)

Errors at output layer nodes can be calculated as follows:

$$e_i = (y_i - T_i) \tag{1.8}$$

Note that errors can be positive or negative; to account for this, their absolute values are considered, i.e.

$$e_i = |y_i - T_i| \tag{1.9}$$

or as squared value

$$e_i^2 = |y_i - T_i|^2 \tag{1.10}$$

It can be represented as Energy Function as a whole:

$$E = \frac{1}{2}\sum_{i} e_i^2 \tag{1.11}$$

For learning procedure, to minimize the error E and to do so they are shown as

$$\frac{\partial E}{\partial W} \text{ or } \frac{\partial E}{\partial t} \tag{1.12}$$

So, the optimization to adjust the weight factors is achieved through a learning algorithm. In this context, the most popular learning algorithm is the 'Back Propagation Algorithm' for Feed Forward NNs, which involves the flow of information from the input layer to the output layer.

It's essential to note that the term 'Back Propagation' is not related to feedback, as information flows from input to output. In conventional control systems, feedback information flows from output to input. However, in this case, it involves error propagation backward, which is related to the adjustment of weights based on the backpropagation of error. Figure 1.7 illustrates the flow of information and error.

Therefore, the information defined between the input and output layers is stored in the weight factors, and they are optimized using the error propagation algorithm. Ultimately, the relationship between the input and output pairs is established through a learning algorithm, without relying on mathematical functional relationships as used in classical system theory.

Note: In the conventional case, the functional relationship $f(\cdot)$ between the dependent and independent variables is given by the equation $y = f(x)$. The output y can be easily found corresponding to input x. However, when the functional relationship $f(\cdot)$ is unknown, the usage of the NN approach becomes more practical.

During the training of the network, sometimes, to achieve enhanced performance, a low noise term is introduced as an additional factor to the data, i.e.

$$\tilde{x}' = \tilde{x} + \text{Noise} \tag{1.13}$$

$$\tilde{y}' = \tilde{y} + \text{Noise} \tag{1.14}$$

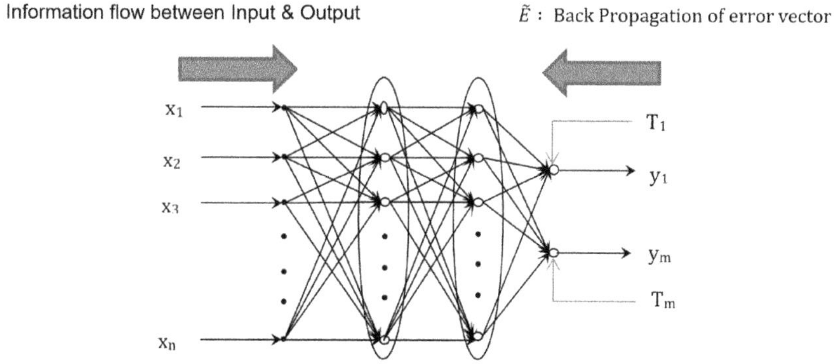

FIGURE 1.7 Feedforward information and error propagation to back in the topology.

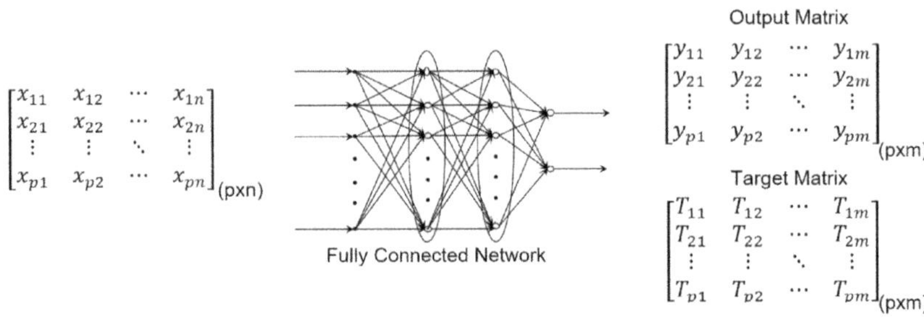

FIGURE 1.8 Matrix notation and topology.

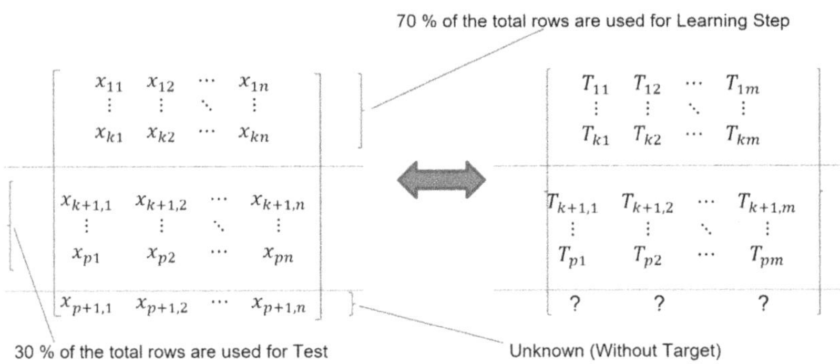

FIGURE 1.9 Data structure for NN.

Considering the noise term as Gaussian Noise[1] with zero mean, unit standard deviation, and gain factor k (usually a small number), noise-addition data is represented as

$$\tilde{x}' = \tilde{x} + k * N(0,1) \tag{1.15}$$

$$\tilde{y}' = \tilde{y} + k * N(0,1) \tag{1.16}$$

The matrix notation of the NN structure can be organized into three different matrixes for the training procedure. These are the input matrix, output matrix, and target matrix, as shown in Figure 1.8.

So by backpropagation of the error matrix and hence adjusting the inner factors, namely the weight factors, actual output will approach the target values at the end of the learning procedure.

1.3.3 The Data Structure Used in the NN Applications

As seen in Figure 1.9, 30% of the input and target data is used for the test procedure, while 70% of them is used as the training set.

NOTE

1 See Appendix A for Gaussian Noise.

2 Artificial Neuron and Its Mathematical Model

2.1 BIOLOGICAL NEURON

A neuron is a fundamental unit of the human brain, facilitating numerous synaptic communications among other neurons. The brain, in turn, acts as an information-processing machine. Each neuron receives multiple inputs through synaptic connections and provides a single output, known as an axon, which carries information to subsequent neurons.

As shown in Figure 2.1, a typical neuron consists of three main parts: the soma (cell body), dendrites, and the axon. Dendrites usually extend their branches, with some reaching several hundred micrometers away from the soma. In contrast, the axon can cover great distances, giving rise to numerous branches.

2.1.1 Neurotransmission

During neurotransmission or synaptic transmission, neurons communicate by transmitting an electrical charge along the axon. These signals are conveyed using chemical messengers known as neurotransmitters, as illustrated in Figure 2.1. The synaptic gap serves as the medium through which the message is passed to the next neuron. Synaptic communication encompasses both electrical and chemical signaling. In electrical synapses, two neurons communicate with each other through a gap junction.

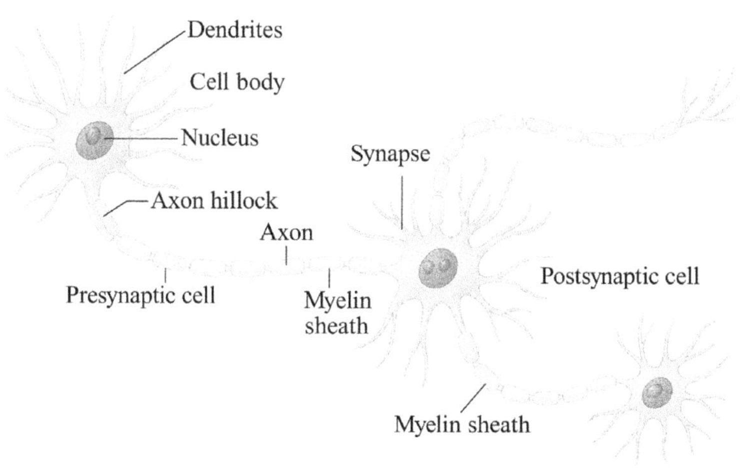

FIGURE 2.1 Basic biological neuron structure.

DOI: 10.1201/9781003673972-2

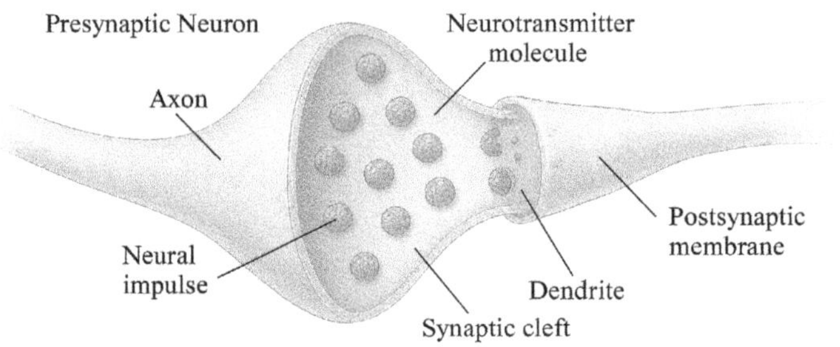

FIGURE 2.2 Enlarged view of the synaptic connection between two neurons.

2.1.2 Electrical Neurotransmission

The communication between two neurons takes place at electrical synapses, often referred to as gap junctions. These synapses enable the exchange of electrical properties between neurons, a process known as synaptic communication or electrical neurotransmission (Figure 2.2).

2.1.3 Chemical Neurotransmission

In chemical neurotransmission between neurons, a gap separates the presynaptic neuron from the postsynaptic neuron. This synaptic cleft is filled with fluid, which is present throughout the brain's cellular environment. Within this nanometer-scale synaptic gap, physical resistance is encountered by electrical charges attempting to travel from one neuron to another. In electrical terms, the synaptic gap can be likened to a 'short' in an electrical circuit. Neurotransmitters serve as the connectors for this electrical 'short circuit' event, functioning as chemical messengers.

Since the information content during neurotransmission exhibits a non-linear structure, the mathematical model of a neuron can be described as follows.

2.2 MATHEMATICAL MODEL OF A BIOLOGICAL NEURON

A neuron is a Multi-Input and Single-Output system, as shown in Figure 2.3, where $[x_1, x_2, \ldots, x_n]$ and $[y_j]$ are the discrete $(n \times 1)$ inputs and output, respectively.

Mathematically, it involves two operations. One of them is the linear combination of the input data, and the other is activation, which represents the actual output. These operations can be represented as follows:

$$a_j = \sum_{i=1}^{n} w_{ji} x_i \tag{2.1}$$

$$y_j = f(a_j) \tag{2.2}$$

Here, $f(\cdot)$ is a non-linear function. There are various types of non-linear activation functions in neural computation, e.g.

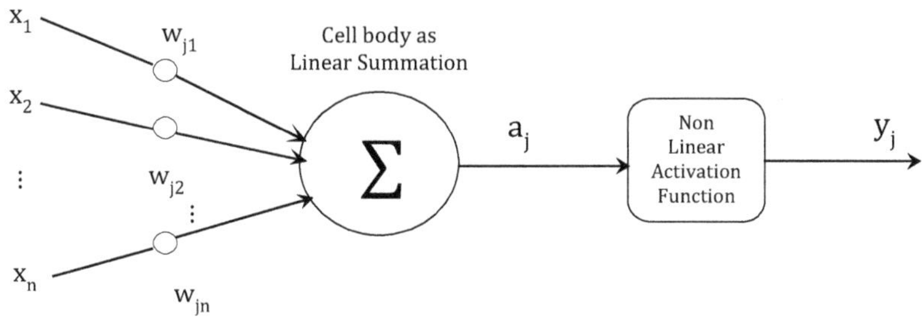

FIGURE 2.3 Artificial model for the *j*th neuron in a neural net.

- 'arctan' function;
- 'tanh' function;
- 'sigmoidal' function.

However, as shown in Figure 2.3, these factors alone are insufficient for the artificial neuron to function at the level of a biological neuron. Therefore, an adaptive algorithm must be employed to adjust the weight factors of this structure. As a result, the learning component becomes an integral part of the entire system, and a new mathematical model can be introduced as a learning unit. This unit is an artificial neuron designed for use in the neural network, featuring individual learning capabilities.

2.2.1 Modified Mathematical Model (as a Learning Unit)

The new mathematical structure of the artificial neuron is shown in Figure 2.4.

Here,

$$x_0 \triangleq -1$$
$$w_{j0} \triangleq \theta_j \text{ (Threshold Value)}$$

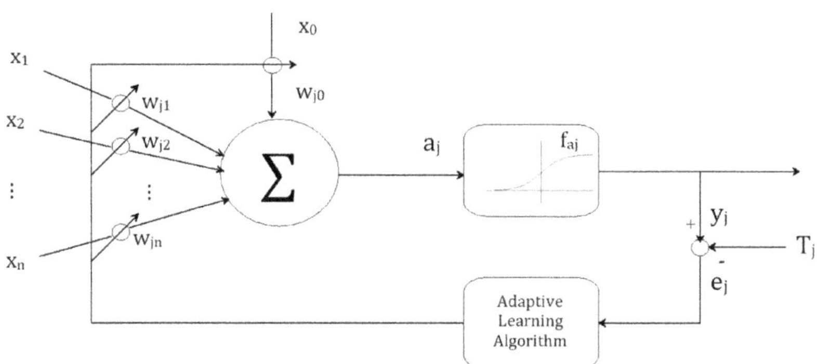

FIGURE 2.4 A dynamical mathematical model of the *j*th neuron by an adaptive learning unit.

w_{ji} : Wieght factors $(i = 1,\ 2,\ \dots, n)$
a_j : Linear output
y_j : Actual output
T_j : Target Value
e_j : error between output & the Target Value (desired output)

$$y_j = f(a_j) = \frac{1}{1 + e^{-a_j}} \tag{2.3}$$

Also, it is known that a_j is the linear combination of weight factors, and hence, the linear sum is calculated as

$$a_j = \sum_{i=0}^{n} w_{ji} x_i \tag{2.4}$$

or it is re-written as

$$a_j = \sum_{i=1}^{n} w_{ji} x_i + w_{j0} x_0 \tag{2.5}$$

Finally,

$$a_j = \sum_{i=1}^{n} w_{ji} x_i + \theta_j(-1) = \sum_{i=1}^{n} w_{ji} x_i - \theta_j \tag{2.6}$$

And furthermore, upon substituting Equation (2.6) into the sigmoidal function, a result is obtained as follows:

$$y_j = \frac{1}{1 + e^{-\sum_{i=1}^{n}(w_{ji} x_i - \theta_j)}} \tag{2.7}$$

or

$$y_j = \frac{1}{1 + e^{\sum_{i=1}^{n}(-w_{ji} x_i + \theta_j)}} \tag{2.8}$$

Thus, the graphical illustration of the sigmoidal activation function used in Equation (2.8) is presented as follows.

In Figure 2.5, for large values of the weights, the output of the activation function asymptotically approaches the unit value. Conversely, for small values of weights, the input is mapped to the region near zero. For this reason, for small values of the weights around the origin, the variation of the activation function can be considered as a linear variation within a very narrow range, as shown in Figure 2.5. Consequently, it can be inferred that for large values of weights, 'the information

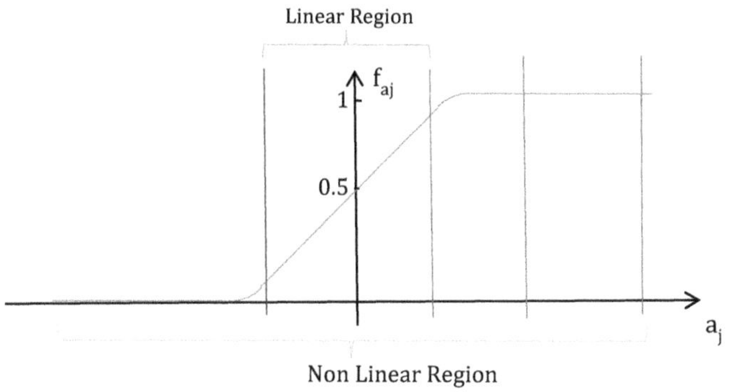

FIGURE 2.5 The logistic (Sigmoidal) function response to the linear sum.

is stored in the non-linearity.' This is a crucial result related to the neurons and their non-linear properties found in nature. Therefore, the stored information can be retained and developed, contributing to the evolution of nature and the learning capacity of the neural network, as discussed in this book on a broader scale.

Therefore, based on the explanations above, it can be concluded that the behavior of the neuron can be adjusted by operating in both the linear and non-linear regions. However, its non-linear nature plays a vital role in this process because information is stored in the non-linearity. This is also a fundamental property of nature. Therefore, as understood, learning and information processing occur within the non-linearity.

2.2.2 Probabilistic Behavior of the Artificial Neuron

A neuron is accepted as a Multi-Input and Single-Output system, and considering its input vector, $X=[x_1, x_2, \ldots, x_n]$, with random variables in the standard normal distribution, i.e., $N(\mu,\sigma^2) \rightarrow N(0,1)$. Hence, the statistical characterization of the neuron response y_j can be defined (Figure 2.6).

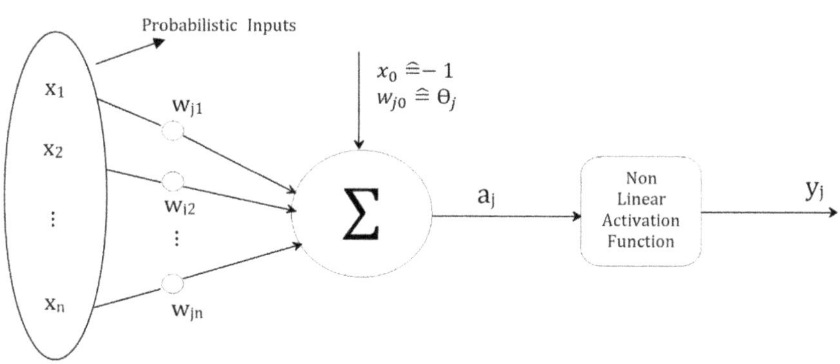

FIGURE 2.6 The response of an artificial neuron to probabilistic inputs.

As shown in Figure 2.5, the non-linear activation function can be divided into linear and non-linear regions of operation. Both regions exhibit different statistical responses to the input vector in terms of standard normally distributed random variables. This is reflected in the probability distribution function due to the non-linear behavior.

2.2.3 Normal Distributed Variables

Consider random variables with a standard normal distribution, where the mean value ($\mu = 0$), standard deviation ($\sigma = 1$), skewness ($s = 0$), and kurtosis ($k = 3$) are shown in Figure 2.7.

In symmetrical distributions, skewness is always zero, while in asymmetrical distributions, skewness takes on non-zero values. It can be observed in Figure 2.8 in the form of left-skewed (for $s < 0$) and right-skewed (for $s > 0$) distributions.

The skewness parameter, denoted as 's,' measures the symmetry of a given normal distribution. In symmetrical cases, it is always zero, while it takes a non-zero value for any asymmetrical distribution.

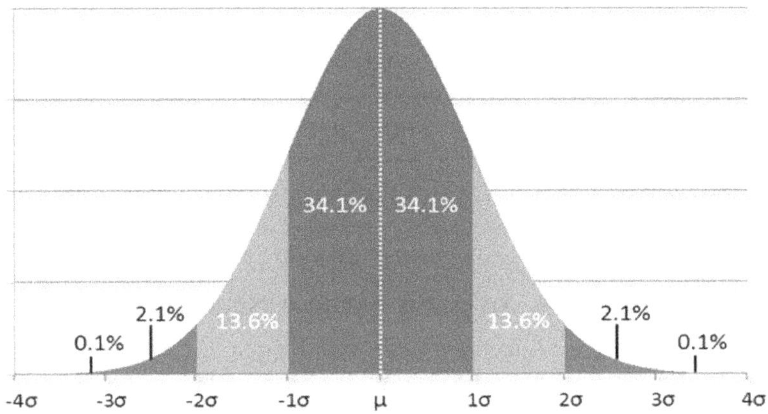

FIGURE 2.7 Standard normal distribution. (https://news.mit.edu/2012/explained-sigma-0209.)

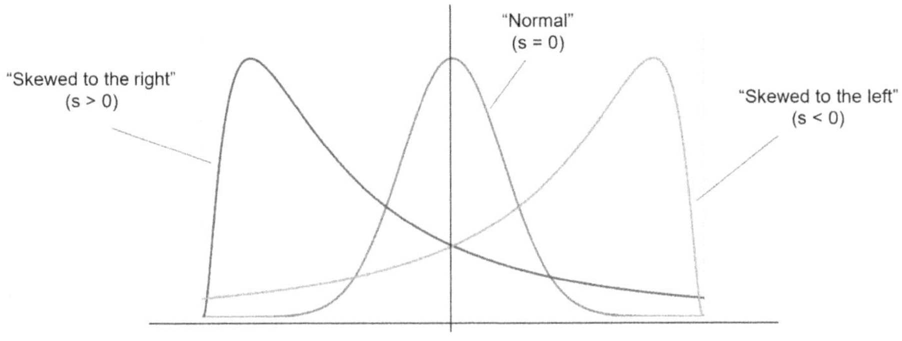

FIGURE 2.8 Symmetrical and asymmetrical distributions.

2.2.4 Response of Linear Transform on the Normal Distributed Variables

The random variables having a standard normal distribution are transformed to another normal distribution with different statistical parameters like

$$Y = WX + \theta \tag{2.9}$$

If an artificial neuron can be linearly modeled by Equation (2.9) and is subsequently passed through the sigmoidal function $f(\cdot)$:

$$y = f(Y) \tag{2.10}$$

It can be concluded that if X follows a standard normal distribution as $\mathcal{N}(0,1)$, with a mean of zero and a standard deviation of one, then, under a linear transformation, the new variables Y will also have a normal distribution, $\mathcal{N}(b,a)$ or $\mathcal{N}(\theta, w^2)$, with different parameters while remaining in the same probability distribution type. However, although the statistical parameters change based on the linear equation parameters, the distribution type remains unchanged, resulting in a Gaussian output for a Gaussian input. This property is not observed in non-linear operations.

However, it can be observed from Figure 2.8 that the linear part of the sigmoidal function is related to the small weight factors on an axis, $\sum_{i}^{a_n} w_{ji}x_i$.

Hence, the linear region of the sigmoidal function transforms the normally distributed (Gaussian) random variables back into a normal distribution. This is depicted in Figure 2.9 for the linear region of the sigmoidal function.

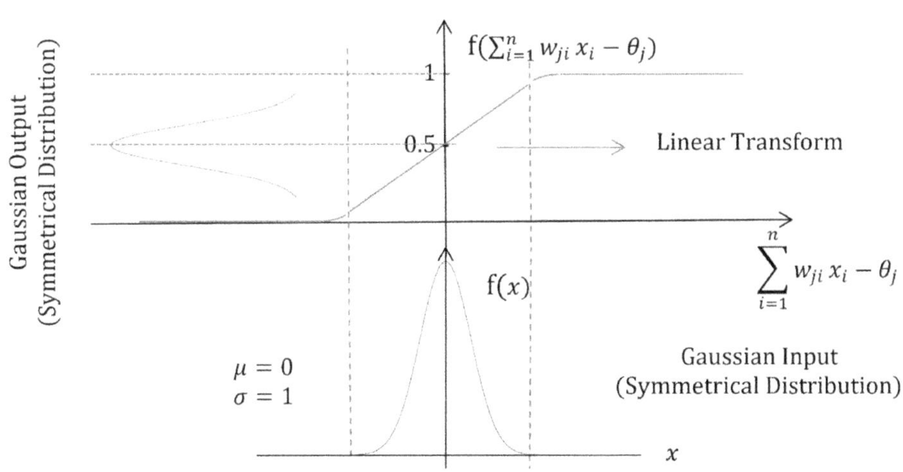

FIGURE 2.9 The linear region response of the sigmoidal function to the standard normal distribution.

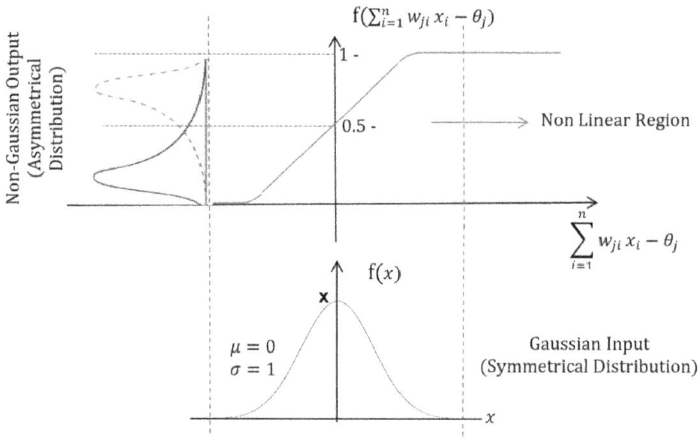

FIGURE 2.10 The non-linear region response of the sigmoidal function to the standard normal distribution.

2.2.5 Response of Non-Linear Transform on the Normal Distributed Variables

Finally, when considering the non-linear transformation of normally distributed random variables within the neuron, as previously concluded, the symmetrical distribution observed in linear transformations is absent. The result of the non-linear transformation leads to an asymmetrical distribution, as depicted in Figure 2.10.

2.3 NON-LINEAR ACTIVATION FUNCTION TYPES

Until now, it has been considered a sigmoidal activation function along with its properties. But other activation functions can also be used as follows. There are (Figures 2.11 and 2.12):

2.3.1 Arctangent Function

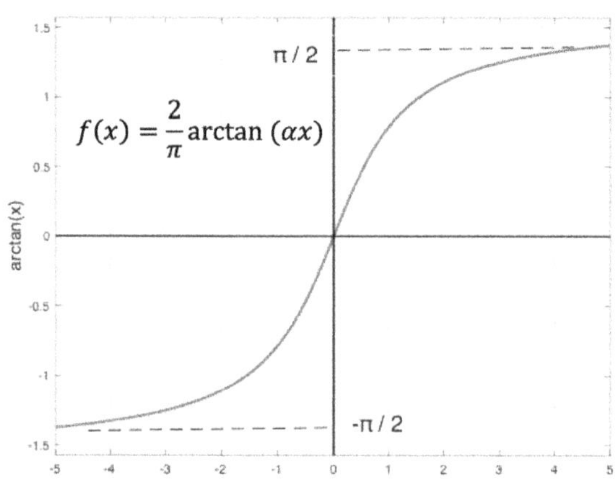

FIGURE 2.11 Arctan function.

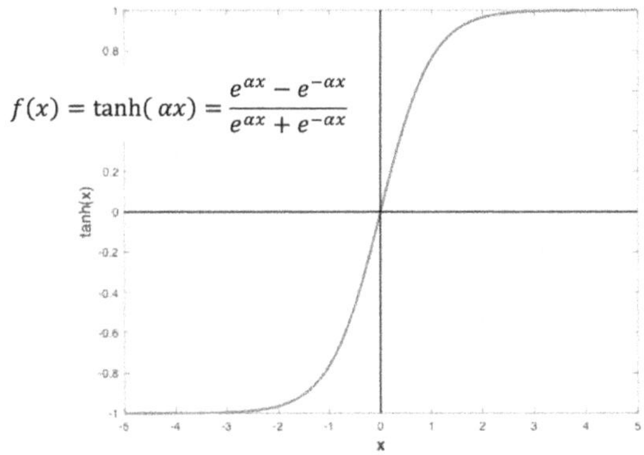

FIGURE 2.12 Hyperbolic tangent function.

2.3.2 Hyperbolic Tangent Function

See Figure 2.12.

2.3.3 Logistic (Sigmoidal) Function with Different Parameters

The logistic function (Sigmoidal function) shows the different variations depending on its slope parameter 'α' as shown in Figure 2.13.

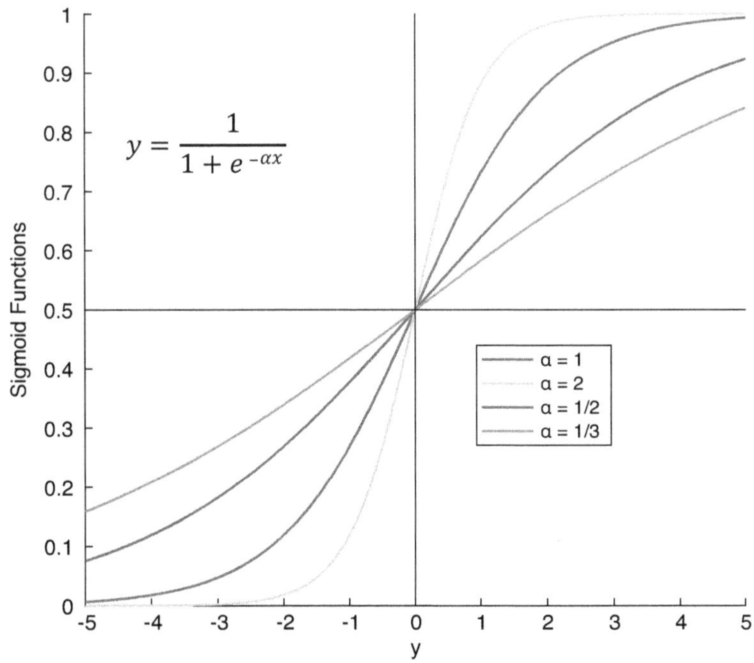

FIGURE 2.13 The sigmoidal function variations with different slope parameter '*a*.'

2.4 ADAPTIVE LEARNING ALGORITHM

A neuron necessitates an adaptive learning process to dynamically adjust its weight factors in accordance with the target (desired) values. This methodology is implemented by Widrow's algorithm.

2.4.1 Widrow-Hopf Delta Learning Rule

Consider a neuron without an activation function but with a target value 'T' and the error value of 'ε:'

If the gradient of the squared error is calculated as follows:

$$\frac{\partial \varepsilon^2}{\partial w_i} = -2(T - Y)x_i \tag{2.11}$$

For two inputs, namely, for w_1 and w_2, the squared error is written as

$$\varepsilon^2 = \left[T - w_1x_1 - w_2x_2\right]^2 \tag{2.12}$$

$$\varepsilon^2 = w_2{}^2\left[x_2{}^2\right] + w_2\left[-2x_2\left(T - w_1x_1\right)\right] + \left[\left(T - w_1x_1\right)^2\right]$$

Hence, to find the minimum point on the plane of (w_1 and w_2), the derivatives are

$$\frac{\partial \varepsilon^2}{\partial w_1} = -2\left[T - w_1x_1 - w_2x_2\right]x_1 = 0 \tag{2.13}$$

$$\frac{\partial \varepsilon^2}{\partial w_2} = -2\left[T - w_1x_1 - w_2x_2\right]x_2 = 0 \tag{2.14}$$

Suppose, $x_1 = x_2 \neq 0$; hence, it becomes $\left[T - w_1x_1 - w_2x_2\right] = 0$. From this equality, w_1 and w_2 are solved, and their optimal values are determined as follows:

$$w_1^* = \frac{T - w_2x_2}{x_1} \tag{2.15}$$

$$w_2^* = \frac{T - w_1x_1}{x_2} \tag{2.16}$$

By substituting the optimal weights w_1^* and w_2^* into Equation (2.12), it yields a Minimum Square Error of zero. This represents the theoretical outcome. However, it is essential to note that in practical, real-world scenarios, the Minimum Square Error does not equal zero due to factors such as non-linearity, noise, and imperfect data.

The minimization of squared error through the Widrow-Hopf training algorithm can be visually demonstrated as depicted in Figure 2.14.

Hence, the Widrow-Hopf rule provides the following equality in terms of the change in each weight:

$$\Delta w_i = -\frac{k\,\partial \varepsilon^2}{\partial w_i} \tag{2.17}$$

or

$$\Delta w_i = 2\mathrm{k}(T-a)x_i = 2\mathrm{k}\varepsilon x_i \tag{2.18}$$

where k is a constant. This rule is known as the delta rule. It follows the gradient of the curve surface toward the optimal weights and is called the 'Gradient Descent' or 'Steepest Descent' Algorithm. Graphically, it is shown in Figure 2.15.

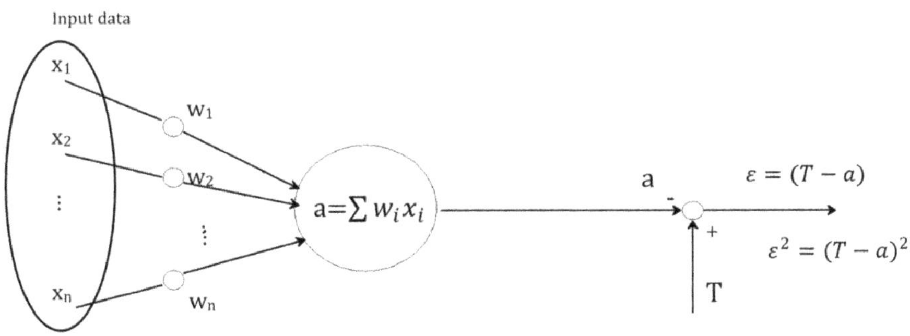

FIGURE 2.14 The learning of the Linear Sum part of the neuron with target value T.

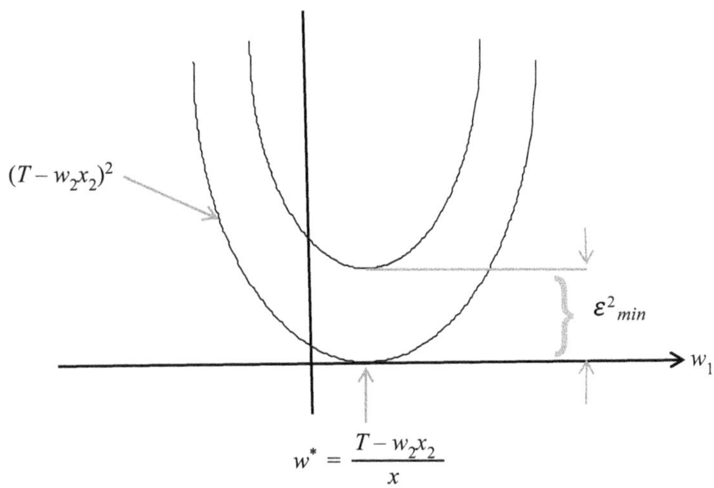

FIGURE 2.15 Gradient Descent Algorithm.

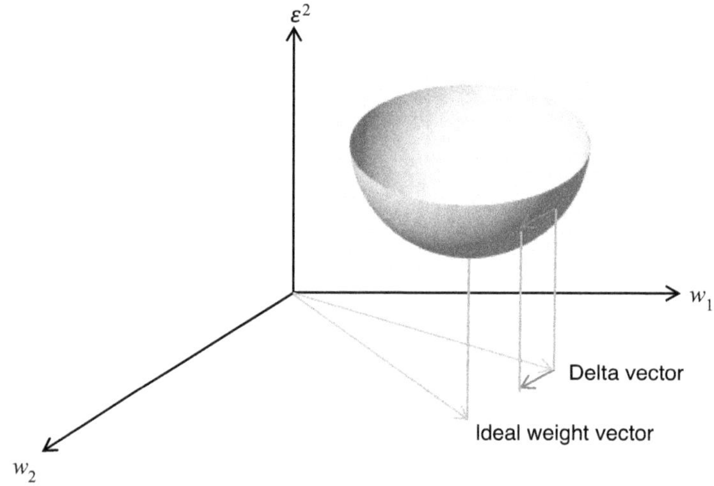

FIGURE 2.16 Geometric representation of the delta rule.

Finally, to normalize the input vector component 'x_i' by $|x_i|^2$, the delta rule is written as (Figure 2.16)

$$\Delta w_i = 2k\varepsilon x_i \frac{|x|^2}{|x|^2} \tag{2.19}$$

$$\Delta w_i = \left[2k|x|^2\right]\frac{\varepsilon x_i}{|x|^2} \tag{2.20}$$

and taking $\eta \cong 2k|x|^2$, it becomes:

$$\Delta w_i = \eta \frac{\varepsilon x_i}{|x|^2} \tag{2.21}$$

Here, the parameter η is named as a learning constant, and it is used in the learning algorithm.

3 Information Flow and Information Content of NN Structure

3.1 SHANNON'S INFORMATION FOR FEEDFORWARD NEURAL NETWORK AND INFORMATION MEASURE AS THE NUMBER OF HIDDEN NODES

Claude Shannon defined the mathematical foundations of information theory in 1948 through a written article. This seminal work introduced the probabilistic definition of entropy.

To understand the concept of entropy as it pertains to information content, it is essential to begin by considering the following two fundamental concepts:

- Information: $I(X)$
- Entropy: $H(X)$

where X is a random vector with probability values of $\left[p(x_1), p(x_2), \ldots, p(x_n)\right]$.

Consequently, the connection between entropy and information concepts can be expressed through the following equations:

$$H(X) = E(I(X)) \tag{3.1}$$

where $E(\cdot)$ is a mathematical operator to calculate the mean value. As a result, this entropy value becomes

$$H(X) = E(-\log_b P(X)) \tag{3.2}$$

where $p(X)$ is the probability mass function of X.

By taking the finite samples, the entropy, $H(X)$ can be written as

$$H(X) = \sum_i p(x_i) I(x_i) \tag{3.3}$$

$$H(X) = -\sum_i p(x_i) \log_b p(x_i) \tag{3.4}$$

DOI: 10.1201/9781003673972-3

where b is the base of the logarithm for the binary case, i.e., $b=2$.

Hence, the mean information or entropy for the binary case becomes

$$H(X) = -\sum_i p_i \log_2 p_i \tag{3.5}$$

Here, $p_i \triangleq p(x_i)$ and $i = 1, \ldots, n$.

Considering the equally likely events defined in the probability theory, each probability value for states n is given as

$$p_i = \frac{1}{n} \tag{3.6}$$

If so, Equation (3.5) becomes

$$H(X) = \sum_{i=1}^{n} p_i \log_2 \frac{1}{p_i} \tag{3.7}$$

and

$$\begin{aligned} H(X) &= \sum_{i=1}^{n} \frac{1}{n} \log_2 (n) \\ &= n\left(\frac{1}{n} \log_2 (n)\right) \\ &= \log_2 (n) \end{aligned} \tag{3.8}$$

For the combination of the entropy (information) sets, e.g., $\{H(X) \text{ and } H(Y)\}$, the conditional entropy is given as (Figure 3.1):

$$\begin{aligned} H(Y|X) &= H(X,Y) - H(X) \\ H(X|Y) &= H(X,Y) - H(Y) \end{aligned} \tag{3.9}$$

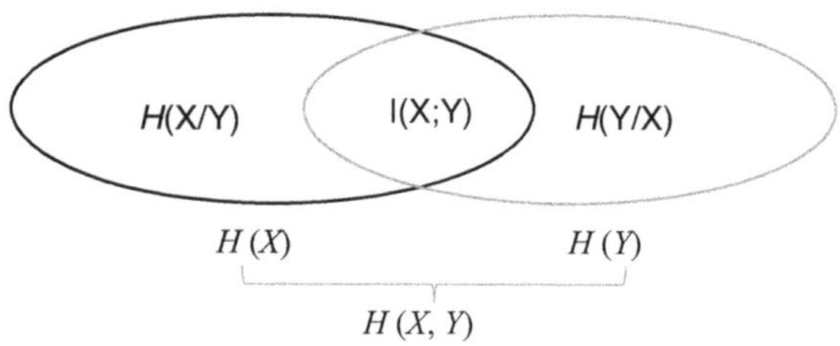

FIGURE 3.1 The Venn diagram of the entropy sets is defined between X and Y pairs.

Information can be understood as the intersection region of the sets in the Venn diagram, and this is formally represented by Equation (3.10):

$$I(X;Y) = H(X) - H\,(X/Y) \tag{3.10}$$

3.2 INTERPRETATION OF A FEEDFORWARD NEURAL NETWORK AS A COMMUNICATION SYSTEM AND SHANNON'S CHANNEL CAPACITY

A fundamental communication system consists of three key components: a transmitter, a receiver, and an information channel. To illustrate, consider the following example.

In the context of the system described above, channel capacity, serving as a measure of information, can be defined using Shannon's approach through the following formula:

$$C = B\ \log_2(1 + \mathrm{SNR}) \tag{3.11}$$

where

C: Channel capacity $[C] =$ bit/sec.
B: Channel Bandwidth $[B] =$ Hz.
SNR: Signal (S) to Noise (N) Ratio.

For a noiseless channel, namely, $N = 0$ and $C = \infty$.

By leveraging the structure (similarity) of the communication channel, as illustrated in Figure 3.2, it can be conceptualized as an FFNN topology. However, to achieve this, several assumptions are made.

We consider a three-layer FFNN that uses only one hidden layer, as shown in Figure 3.3.

As a result, the hidden layer can be likened to the concept of channel capacity. Moreover, the numerical value of the channel capacity can be characterized in

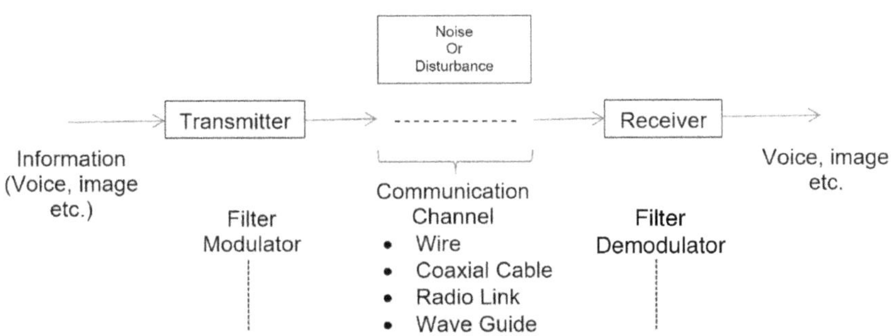

FIGURE 3.2 Basic communication system.

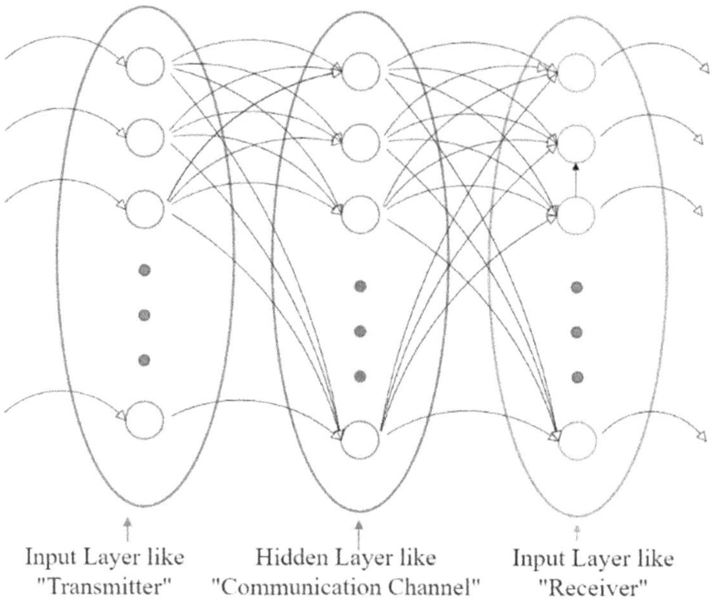

FIGURE 3.3 An FFNN network topology presenting the communication system.

terms of the number of hidden units, much like the representation of information in bits. Consequently, the neural network's topology can be framed as an optimization problem aimed at determining the optimal number of hidden nodes within a single hidden layer.

3.3 INTERPRETATION OF NEURAL NETWORKS AS COMMUNICATION SYSTEMS

In the context of an FFNN topology, the challenge of determining the number of hidden nodes in the hidden layer remains an open problem. However, in this chapter, when considering a single hidden layer situated between the input and output layers, the number of hidden nodes can be quantified using an information measure. This information measure is founded on Shannon's information theory.

Consequently, it can be characterized as a logarithmic measure denominated in bits. The entropy value, which serves as a quantification of the information, can be expressed for a binary state as follows:

$$H(p_j,q_j) = -p_j\log_2(p_j) - q_j\log_2(q_j) \tag{3.12}$$

$$H(p_j, 1-p_j) = -p_j\log_2(p_j) - (1-p_j)\log_2(1-p_j) \tag{3.13}$$

Also, the graphical illustration of the entropy function $H_j \triangleq H(p_j,q_j)$ for the binary state becomes (Figure 3.4):

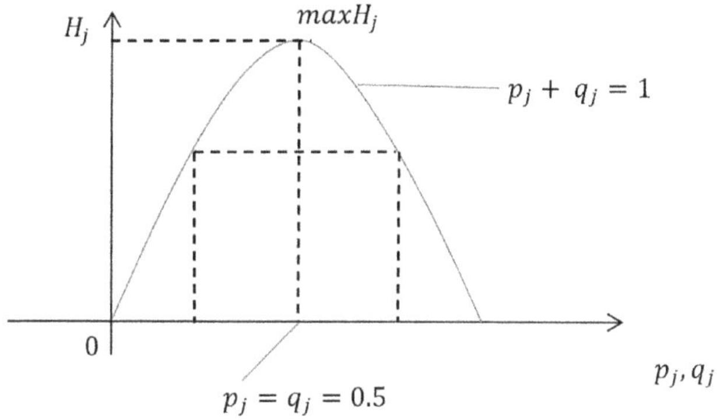

FIGURE 3.4 The variation of the binary entropy function.

In certain cases, rather than using binary states, the representation can be expressed in terms of the natural logarithm with a base of 'e.' In such cases, Equation (3.11) can be reformulated as

$$H(p_j, 1-p_j) = -p_j \ln(p_j) - (1-p_j)\ln(1-p_j) \tag{3.14}$$

As indicated in Equation (3.12), the entropy function involves the use of probability values. However, in a neural network topology, there are typically no inherent probability values associated with the data. To impart a probabilistic interpretation to neural network applications, certain assumptions can be taken into account:

3.3.1 Assumptions and Definitions

When a pattern set is applied to the neural network topology, the connection between input and output pairs can be described by a function f_0, which is termed the mapping function between the input and output vector spaces. This function relates the input patterns to their corresponding output patterns:

$$\tilde{X} \in R^N,\ \tilde{Y} \in R^M \tag{3.15}$$

$$f_0 : A \subset R^N \Rightarrow R^M \tag{3.16}$$

Indicating the states of each independent input vector by index 's,' it can be shown as in Figure 3.5.

Here, weight matrixes, $\left[W_{ij}^*\right]$ and $\left[W_{jk}^*\right]$, are constant matrixes after the training procedure. And also, taking $N=K$, it can be assumed as an auto-associative neural net.

If we choose each pattern vector s from the pattern set A with dimensions of $N \times N$, then the probability value of each pattern vector p(s) can be computed as uniformly distributed, as demonstrated below:

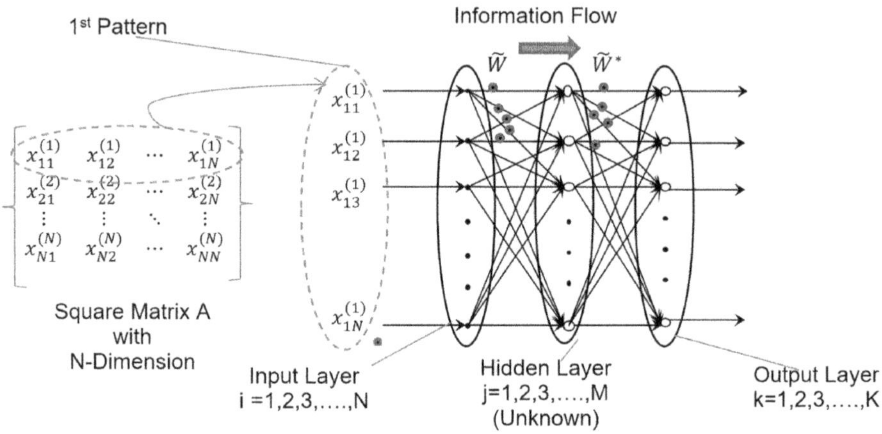

FIGURE 3.5 The information flows in the FFNN topology with three layers. [(*) Auto-Associative NN.]

$$p(s) \triangleq \frac{1}{N} \tag{3.17}$$

Let's consider the information flow from the input layer to the hidden layer nodes. The weighted sum of the jth processing element of the hidden layer at state (s) is conveyed to the output layer via the sigmoidal function. Consequently,

$$y_j^{(s)} = \frac{1}{\left\{1+\exp\left[-\beta\left(\sum W_{ji}{}^{*} x_i^{(s)} - \theta_j\right)\right]\right\}} \tag{3.18}$$

In the equation, the optimal weights acquired after training are represented. In this context, the half topology of the FFNN is employed to depict the information content within the single hidden layer, which is portrayed as input and hidden output pairs. Consequently, the hidden output node matrix becomes $\tilde{Y}$ and its elements $y_j \in \tilde{Y}$.

Hence, the exponential term is given by

$$\beta\left(\sum_i W_{ji}{}^{*} x_i^{(s)} - \theta_j\right) = \ln\left(\frac{y_j^{(s)}}{1 - y_j^{(s)}}\right) \tag{3.19}$$

This exponential solution can be comprehended in terms of two facets of the equation. The right side, expressed in logarithmic form, signifies the amount of information. Simultaneously, the left side reveals the information matrix established at the hidden layer nodes through the utilization of the coefficient matrix defined between the input and hidden layer nodes. The subsequent section will provide a more dynamic exposition of these interpretations, employing concepts related to entropy functions.

3.3.2 Application of the Entropy Function

Using the definition of the binary entropy function, it is given as

$$H(p_j, 1-p_j) = -p_j \ln(p_j) - (1-p_j)\ln(1-p_j). \tag{3.20}$$

In this context, the entropy definition can be reformulated as follows for the exponential term of the sigmoidal outputs at the hidden nodes, as presented in Equation (3.19). Consequently, when considering Equation (3.18), it becomes

$$\beta\left(\sum_{i,s} W_{ji}^{*} p_i(s) - \theta_j\right) = \frac{-dH}{dp_j} \tag{3.21}$$

where

$$x_i^{(s)} \rightarrow p_i(s),\ y_j^{(s)} \rightarrow p_j(s),\ \text{and}\ \frac{-dH}{dp_j} = \ln\left(\frac{p_j}{1-p_j}\right)$$

This last equality is the stochastic differential equation for the information flow in the neural network.

By integrating Equation (3.21) and with the assistance of Equation (3.10), we can arrive at the following expression:

$$-\beta\left(\sum_{i,s} W_{ji}^{*} p_i(s) - \theta_j\right) = I(X;Y) \tag{3.22}$$

In the case of auto-associative systems where $I(X,Y) = H(Y)$, if we compare the left and right sides of Equation (3.21), it can be expressed as

$$\beta.\theta_j \triangleq H(X) \tag{3.23}$$

Assuming that all bias terms are equal to each other, taking '$g = 1$,'

$$\beta\left(\sum_{j=1} g\right) = -\left(\frac{1}{N}\right)\ln\left(\frac{1}{N}\right) \tag{3.24}$$

Here, $j = 1, \ldots, M$, and the coefficient μ is used to convert the natural logarithm to binary base:

$$\mu\beta MN = \log_2(N) \tag{3.25}$$

And, taking the new coefficient as $B = 1/(\beta\mu N)$, it is rewritten as

$$M = B\log_2(N) \tag{3.26}$$

Nonetheless, when considering the channel capacity formula established in communication systems, the number of hidden nodes in single-layer FFNNs can be effectively treated as the channel capacity. This can be expressed similarly to Equation (3.27):

$$M' = B\,[\![\log_2(N)]\!] = B(\log_2(N) + 1) \tag{3.27}$$

where

M': The number of hidden nodes with the similarity of channel capacity.
$[\![.]\!]$: Integer function rounding to the upper value, adding +1 to the integer part.

As a result, the FFN topology can be depicted as a communication system, and the quantity of hidden nodes within a single hidden layer can be defined by the channel capacity of the communication system, serving as a measure of information.

4 Backpropagation Algorithm

4.1 BACKPROPAGATION ALGORITHM

The backpropagation algorithm is one of the most renowned and earliest algorithms. It was originally introduced by Werbos in 1974. This algorithm is a type of feedforward neural structure in which the error terms arising at the output nodes are retroactively transmitted back to the input layer from the output layer. This backward propagation of errors facilitates the adjustment of internal parameters or weights of the neural network through the training process. Hence, it is aptly named the backpropagation algorithm.

Let us consider a three-layer feedforward neural network topology, as shown in Figure 4.1.

Where:

1. Input (i) for the 'p' pattern is given as

$$X_{pi} = \left(x_{p1},\ x_{p2}, x_{p3},\ \ldots,\ x_{pN}\right) \text{ Or } \begin{matrix} x_{11} & x_{12} & \cdots & x_{1N} \\ x_{21} & x_{22} & \cdots & x_{2N} \\ \vdots & \vdots & \ddots & \vdots \\ x_{p1} & x_{p2} & \cdots & x_{pN} \end{matrix} \tag{4.1}$$

2. Net input to the hidden layer neurons (j) is given as

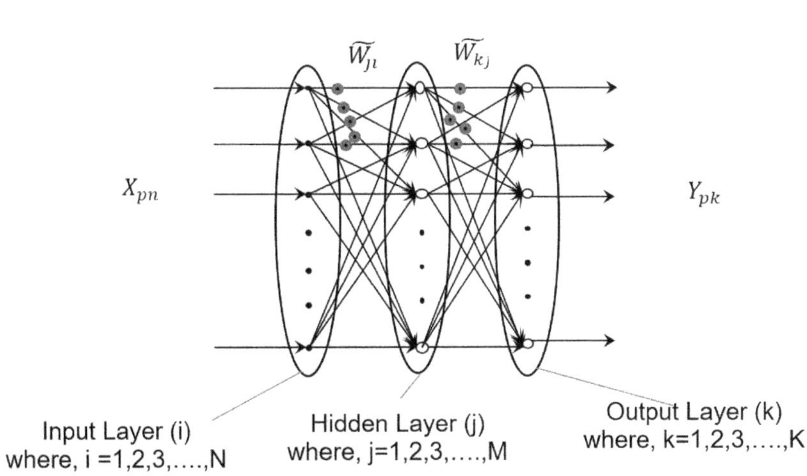

FIGURE 4.1 Multilayered feedforward neural network with one hidden layer.

DOI: 10.1201/9781003673972-4

$$\text{net}_{pj} = \sum_{i=1}^{N} W_{ji} \ X_{pi} \tag{4.2}$$

3. The output of the hidden layer neurons (j) based on the actuation function (e.g., sigmoidal function) is given as

$$\text{Output}_{pj} = f\left(\text{net}_{pj}\right) \triangleq f_{pj} \tag{4.3}$$

4. Net input to the output layer neurons (k) is given as

$$\text{net}_{pk} = \sum_{j=1}^{M} W_{kj} \ \text{Output}_{pj} \tag{4.4}$$

5. The output of the output layer neurons, denoted as k, based on the activation function (e.g., sigmoidal function), can be expressed as follows:

$$\text{Output}_{pk} = f\left(\text{net}_{pk}\right) \triangleq f_{pk} \tag{4.5}$$

4.2 ADJUSTMENT/UPDATING OF WEIGHT FACTORS BETWEEN OUTPUT AND HIDDEN LAYERS

Now, we will establish an energy function for the output nodes of the neural network topology, as illustrated in Figure 4.1. This energy function can be expressed as follows:

$$E_{pk} = \frac{1}{2}\left(t_{pk} - \text{Output}_{pk}\right)^2 \tag{4.6}$$

The main goal of the backpropagation algorithm is to deduce the method/conditions for adjustment of weight factors based on the minimum value of the energy function, E_{pk}, as follows:

$$\frac{\partial E_{pk}}{\partial W_{kj}} = \frac{\partial E_{pk}}{\partial \text{net}_{pk}} \cdot \frac{\partial \text{net}_{pk}}{\partial W_{kj}} \tag{4.7}$$

Using Equation (4.4),

$$\frac{\partial \text{net}_{pk}}{\partial W_{kj}} = \text{Output}_{pj} \tag{4.8}$$

Equation (4.7) becomes

$$\frac{\partial E_{pk}}{\partial W_{kj}} = \frac{\partial E_{pk}}{\partial \text{net}_{pk}} \cdot \text{Output}_{pj} \tag{4.9}$$

Re-writing the first term of the right-hand side of Equation (4.9) using the chain rule:

$$\frac{\partial E_{pk}}{\partial \text{net}_{pk}} = \frac{\partial E_{pk}}{\partial \text{Output}_{pk}} \cdot \frac{\partial \text{Output}_{pk}}{\partial \text{net}_{pk}} \tag{4.10}$$

Using Equations (4.5) and (4.6), we can have the following equations:

$$\frac{\partial \text{Output}_{pk}}{\partial \text{net}_{pk}} = \frac{\partial}{\partial \text{net}_{pk}} f\left(\text{net}_{pk}\right) = f'\left(\text{net}_{pk}\right) = f'_{pk} \tag{4.11}$$

and

$$\frac{\partial E_{pk}}{\partial O_{pk}} = -\left(t_{pk} - \text{Output}_{pk}\right) \tag{4.12}$$

Thus, Equation (4.10) using Equations (4.11) and (4.12) becomes

$$\frac{\partial E_{pk}}{\partial \text{net}_{pk}} = -f'_{pk}\left(t_{pk} - Output_{pk}\right) \tag{4.13}$$

Also, using the following definition,

$$-\frac{\partial E_{pk}}{\partial \text{net}_{pk}} = \delta_{pk} \tag{4.14}$$

We can define

$$\delta_{pk} = f'_{pk}\left(t_{pk} - \text{Output}_{pk}\right) \tag{4.15}$$

Returning to Equation (4.9) and substituting Equation (4.15), it becomes

$$\frac{\partial E_{pk}}{\partial W_{kj}} = -\delta_{pk}\text{Output}_{pj} \tag{4.16}$$

The equation above reveals the decreasing behavior of the energy function in relation to the weight factors associated with the output nodes of the neural topology. Consequently, it can be demonstrated in terms of proportionality, illustrating the connection between alterations in the weights between the hidden and output nodes:

$$\Delta_p W_{kj} \propto \delta_{pk}\text{Output}_{pj} \tag{4.17}$$

Additionally, by incorporating a parameter known as the learning rate (η), this proportionality is equated as follows:

$$\Delta_p W_{kj} = \eta\delta_{pk}\text{Output}_{pj} \tag{4.18}$$

In Equation (4.18), the left-hand side represents the finite difference, and it can be expressed based on the iteration number (n) as follows:

$$W_{kj}(n+1) - W_{kj}(n) = \eta\delta_{pk}\text{Output}_{pj} \tag{4.19}$$

or

$$W_{kj}(n+1) = W_{kj}(n) + \eta\delta_{pk}\text{Output}_{pj} \tag{4.20}$$

For f'_{pk}, we can calculate the derivative of the sigmoidal function (chosen as activation function) as follows:

$$f'_{pk} = \frac{\partial}{\partial \text{net}_{pk}}\left[\frac{1}{1+e^{-\beta \text{net}_{pk}}}\right] \tag{4.21}$$

$$f'_{pk} = \frac{\beta e^{-\beta \text{net}_{pk}}}{\left(1+e^{-\beta \text{net}_{pk}}\right)^2} = \beta \text{Output}_{pj}\left(1-\text{Output}_{pj}\right) \tag{4.22}$$

Using the calculated result of f'_{pk}, Equations (4.15) and (4.20) can be finally re-written as

$$\delta_{pk} = \beta \text{Output}_{pj}\left(1-\text{Output}_{pj}\right)\left(t_{pk} - \text{Output}_{pk}\right) \tag{4.23}$$

$$W_{kj}(n+1) = W_{kj}(n) + \eta \text{Output}_{pj}\left(1-\text{Output}_{pj}\right)\left(t_{pk} - O_{pk}\right)\text{Output}_{pj} \tag{4.24}$$

Here, $\beta = 1$, normally the β parameter can be taken as $0 < \beta \leq 1$.

4.3 ADJUSTMENT/UPDATING OF WEIGHT FACTORS BETWEEN HIDDEN AND INPUT LAYERS

A similar approach can be applied to update the weight factors defined between the hidden and input layers. Consequently, to formulate Equation (4.20) for this new scenario, specifically for W_{ji} and Output_{pi}, we necessitate the value of δ_{pj}. It's noteworthy that δ_{pj} can be conveniently calculated based on δ_{pk}, and this process is fundamental to the backpropagation algorithm.

In this context, we are examining the total energy E_p of the output nodes:

$$E_p = \sum_k E_{pk} \tag{4.25}$$

Then, by using the derivative of the sigmoidal function and the parameters between the hidden and input layers, we can derive the minimum condition as follows:

$$\frac{\partial E_p}{\partial \text{net}_{pj}} = \frac{\partial E_p}{\partial \text{Output}_{pj}} \cdot \frac{\partial \text{Output}_{pj}}{\partial \text{net}_{pj}} \tag{4.26}$$

Equation (4.26) can be reformulated as follows:

$$\frac{\partial E_p}{\partial \text{net}_{pj}} = \sum_k \frac{\partial E_{pk}}{\partial \text{Output}_{pj}} \cdot \frac{\partial \text{Output}_{pj}}{\partial \text{net}_{pj}} \tag{4.27}$$

In the above equation, the first term on the right-hand side can be computed by applying the chain rule:

$$\frac{\partial E_p}{\partial \text{Output}_{pj}} = \sum_k \frac{\partial E_{pk}}{\partial \text{net}_{pk}} \cdot \frac{\partial \text{net}_{pk}}{\partial \text{Output}_{pj}} \tag{4.28}$$

Putting the net_{pk} from Equation (4.2), we have

$$\frac{\partial E_p}{\partial \text{Output}_{pj}} = \sum_k \frac{\partial E_{pk}}{\partial \text{net}_{pk}} \cdot \frac{\partial}{\partial \text{Output}_{pj}} \left(\sum_j W_{kj} \ \text{Output}_{pj} \right) \tag{4.29}$$

or

$$\frac{\partial E_p}{\partial \text{Output}_{pj}} = \sum_k \frac{\partial E_{pk}}{\partial \text{net}_{pk}} \cdot W_{kj} \tag{4.30}$$

Inserting δ_{pk} as we defined in Equation (4.14) in the previous case, we have

$$\frac{\partial E_p}{\partial \text{Output}_{pj}} = -\sum_k \delta_{pk} \cdot W_{kj} \tag{4.31}$$

For the second term on the right-hand side of Equation (4.27), we have

$$\frac{\partial \text{Output}_{pj}}{\partial \text{net}_{pj}} = \frac{\partial}{\partial \text{net}_{pj}} f\left(\text{net}_{pj}\right) = f'\left(\text{net}_{pj}\right) = f'_{pj} \tag{4.32}$$

By applying Equations (4.31) and (4.32), we can express Equation (4.26) as

$$\frac{\partial E_p}{\partial \text{net}_{pj}} = -f'_{pj} \sum_k \delta_{pk} \cdot W_{kj} \tag{4.33}$$

or

$$-\frac{\partial E_p}{\partial \text{net}_{pj}} = f'_{pj} \sum_k \delta_{pk} \cdot W_{kj} \tag{4.34}$$

Using a similar definition of δ_{pk} in Equation (4.14), we can define

$$-\frac{\partial E_p}{\partial \text{net}_{pj}} = \delta_{pj} \tag{4.35}$$

Therefore, Equation (4.33) can be reformulated as

$$\delta_{pj} = f'_{pj} \sum_k \delta_{pk} \cdot W_{kj} \tag{4.36}$$

The equation defined for δ_{pj} can be applied in the computation of $\Delta_p W_{kj}$ which represents the changes in weights W_{kj} between the input and hidden layers, much like in the previous case as described in Equations (4.17) and (4.18). This allows for the adjustment of these weights during the learning process:

$$\Delta_p W_{ji} = \eta \delta_{pj} \text{Output}_{pj} \tag{4.37}$$

In Equation (4.37), the left-hand side is the finite difference, and it can be re-written depending on the iteration number (n) as follows:

$$W_{ji}(n+1) - W_{ji}(n) = \eta \delta_{pj} \text{Output}_{pi} \tag{4.38}$$

or

$$W_{ji}(n+1) = W_{ji}(n) + \eta \delta_{pj} \text{Output}_{pi} \tag{4.39}$$

For f'_{pj}, we can calculate the derivative of the sigmoidal function (chosen as actuation function) as follows:

$$f'_{pj} = \frac{\partial}{\partial \text{net}_{pj}} \left[\frac{1}{1+e^{-\beta \text{net}_{pj}}} \right] \tag{4.40}$$

$$f'_{pj} = \frac{\beta e^{-\beta \text{net}_{pj}}}{\left(1+e^{-\beta \text{net}_{pj}}\right)^2} = \beta \text{Output}_{pj}\left(1 - \text{Output}_{pj}\right) \tag{4.41}$$

Using the calculated result of f'_{pj}, Equations (4.36) and (4.39) can be finally re-written as

$$\delta_{pj} = \text{Output}_{pj}\left(1 - \text{Output}_{pj}\right) \sum_k \delta_{pk} \cdot W_{kj} \tag{4.42}$$

$$W_{kj}(n+1) = W_{kj}(n) + \eta \text{Output}_{pj}\left(1 - \text{Output}_{pj}\right) \sum_k \delta_{pk} \cdot W_{kj} \text{Output}_{pi} \tag{4.43}$$

Here $\beta = 1$, normally the β parameter can be taken as $0 < \beta \leq 1$. And the parameter η $(0 < \eta < 1)$ is the learning rate (or constant). For a good learning level, it can be selected as $\eta = 0.9$. For stability of the learning process, a lower value η for is normally chosen.

4.4 SCHEMATIC REPRESENTATION OF THE BACKPROPAGATION METHOD

Figure 4.2 provides a schematic representation of the backpropagation algorithm, highlighting the key stages of forward information flow and backward error propagation. It visually demonstrates how the output error is computed and distributed back through the network layers to adjust the weights. The diagram serves as a conceptual aid for understanding the learning dynamics within a multilayer feedforward neural network.

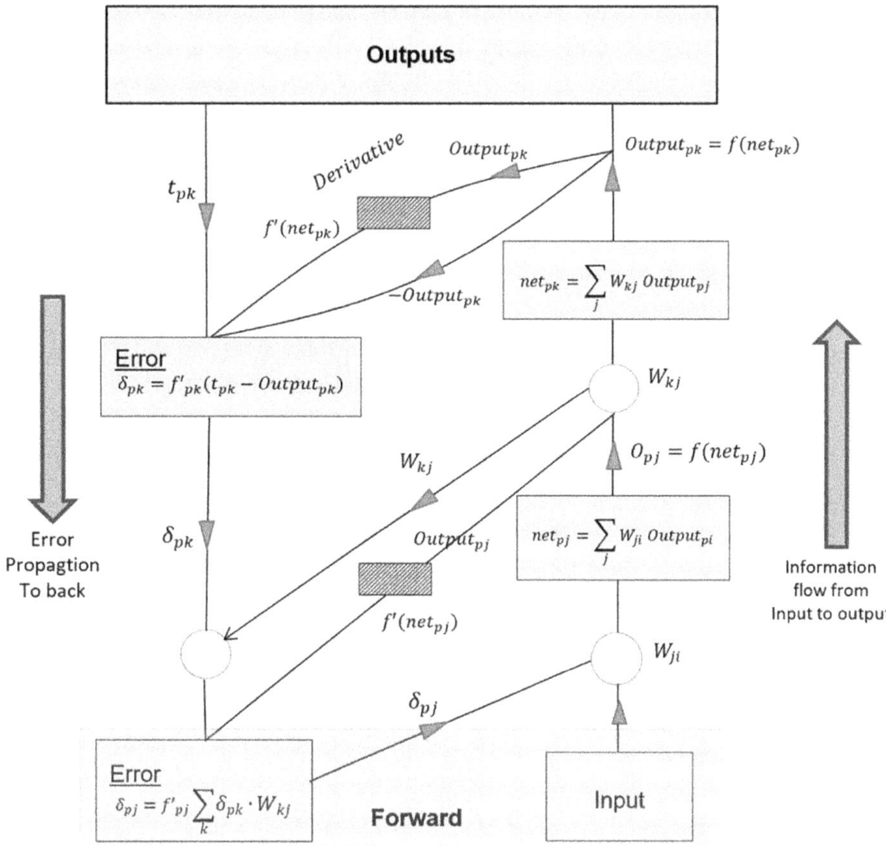

FIGURE 4.2 Schematic flow diagram for the backpropagation algorithm.

4.5 BACKPROPAGATION ALGORITHM AND ITS FLOW CHART

The backpropagation algorithm is a fundamental component of supervised learning used in multilayer feedforward neural networks. Here, the key steps of the algorithm are outlined, from computing the output error to propagating it backward through the layers to update connection weights. The goal is to minimize the overall error by iteratively refining the network's parameters based on the difference between actual and desired outputs. A schematic flow chart accompanies the explanation to provide a clearer understanding of how the learning process unfolds.

4.5.1 Algorithm

1. For all connections, the weight factors among layers can be initialized by small random numbers using a random number generator (like standard normally distributed random numbers).
2. Define the counter of the number of iterations and read its first value as $n = 1$.

3. Define the counter of the number of patterns and read its first value as $p=1$.
4. Read the first pattern and execute it.
5. Using this first pattern, calculate the outputs of neurons by using Equations (4.2)–(4.5).
6. Update the weights between the layers using Equations (4.20) and (4.43).
7. Repeat the calculations between Steps 3–6 up to the end of all pattern sets.
8. For the nth iteration, calculate the error function $E(n)$ as below:

9. $$E(n)=\sum_p E_p=\sum_p \frac{1}{2}\sum_k \left(t_{pk}-\text{Output}_{pk}\right)^2 \tag{4.44}$$

10. Consider a very small number ε, like $0<\varepsilon<1$ and if it provides the following condition for the error function (energy function), go to stop.

11. $$\left|E(n)-E(n-1)\right|<\varepsilon \tag{4.45}$$

12. Save all updated weight factors. In the opposite case, take the next iteration number, go to Step 3, and repeat the operations between Steps 3–9 until reaching the acceptable level numerical value of the error parameter ε.

4.5.2 The Flow Chart of BP-Algorithm

Understanding the operational structure of the backpropagation algorithm is essential for implementing effective learning in multilayer neural networks. This subsection introduces a visual summary of the algorithm's learning cycle to complement the earlier mathematical descriptions. By mapping each phase of the process—initialization, forward propagation, error computation, and weight adjustment—the flow chart provides a clear pathway from input to learning outcome. Such representations are valuable not only for conceptual understanding but also for practical applications in areas such as pattern recognition, control systems, and predictive modeling.

Figure 4.3 illustrates the flow chart of the backpropagation algorithm, providing a clear step-by-step visualization of the learning process. It begins with the initialization of weights, followed by forward propagation, error computation, and iterative weight adjustment. This figure is intended to help readers grasp the dynamic cycle through which the network learns from input–output data until the desired accuracy is achieved. The schematic serves as a practical complement to the mathematical explanation provided in this section.

4.6 APPLICATION: IMPLEMENTING THE BACKPROPAGATION ALGORITHM FOR NONLINEAR FUNCTION APPROXIMATION

Backpropagation is the fundamental algorithm behind the supervised training of Artificial Neural Networks (ANN), enabling them to learn complex, nonlinear relationships between inputs and outputs. The effectiveness of backpropagation lies in its ability to iteratively adjust network weights to minimize prediction error by

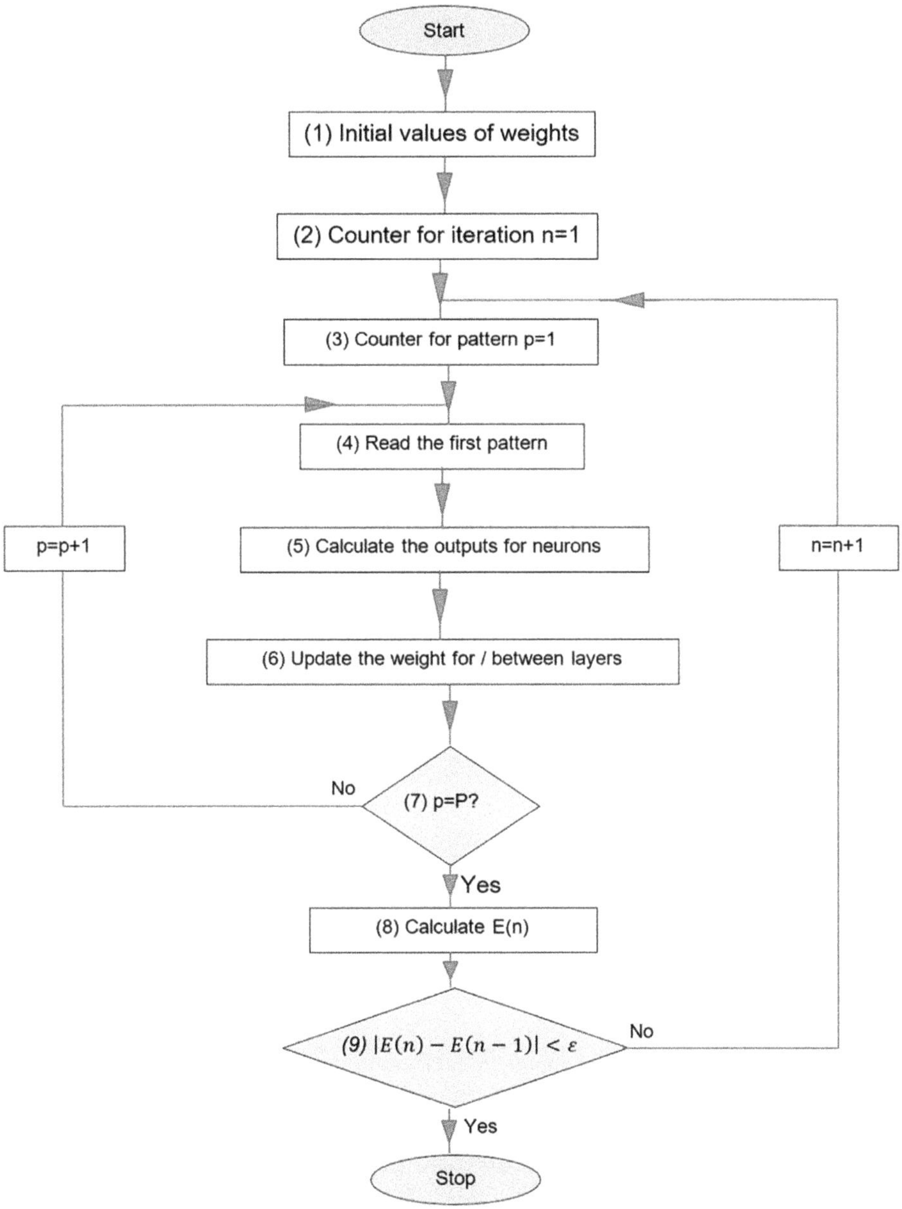

FIGURE 4.3 Flow chart of the backpropagation algorithm.

propagating the gradient of the error backward through the layers. This section demonstrates the practical application of backpropagation in function approximation using a feedforward neural network trained on a synthetic dataset.

In this study, a nonlinear cubic function with added Gaussian noise is used to simulate real-world data variability. The dataset is generated as

$$y = 0.5x^3 - 2x^2 + 3x + \in \quad (4.46)$$

where x is sampled uniformly over [−5, 5] and ϵ represents *zero-mean Gaussian noise* with a standard deviation of 0.2. The objective is to train a neural network to approximate this function, learning its complex dependencies while handling the effects of noise.

A feedforward neural network is employed, consisting of an input layer, a hidden layer with 10 neurons, and an output layer. The network learns by adjusting its weight parameters using the backpropagation algorithm, which minimizes the mean squared error (MSE) between the true values y and the predicted values $\hat{y}$. The weight update rule follows:

$$w_{ij}(t+1) = w_{ji}(t) - \eta \frac{\partial E}{\partial w_{ji}} \quad (4.47)$$

where w_{ij} represents the weight connecting neuron iii to neuron j, η is the learning rate, and E is the total error function defined as

$$E = \frac{1}{N} \sum_{n=1}^{N} (y_n - \hat{y}_n)^2 \quad (4.48)$$

Through iterative training, the network learns to approximate the target function while minimizing the error. The trained model is evaluated based on its generalization ability, tested on previously unseen data. Visualizing the model's performance through prediction plots, error histograms, and regression analysis helps assess the effectiveness of backpropagation in learning complex mappings.

This case study illustrates the strength of ANNs in nonlinear function approximation and provides a structured approach to implementing backpropagation for data-driven modeling.

Figure 4.4 illustrates the architecture of a feedforward neural network utilized for approximating nonlinear functions. The network comprises an input layer with a single neuron, a hidden layer with 10 neurons, and an output layer with 1 neuron. The hidden layer employs a nonlinear activation function to capture complex patterns in the data, while the output layer maps the learned representation to the target output.

Key components of the architecture include:

- Weights (W) and biases or (b) for each layer, which are adjusted during training via backpropagation to minimize error.
- Activation functions, indicated by the nonlinear symbols, applied to the weighted sums of inputs to introduce nonlinearity.

This network design demonstrates the ability of feedforward neural networks to model complex relationships, such as those in the synthetic dataset used in this chapter.

Figure 4.5 depicts the synthetic dataset used for training a feedforward neural network to approximate a nonlinear function. The marker points represent the noisy

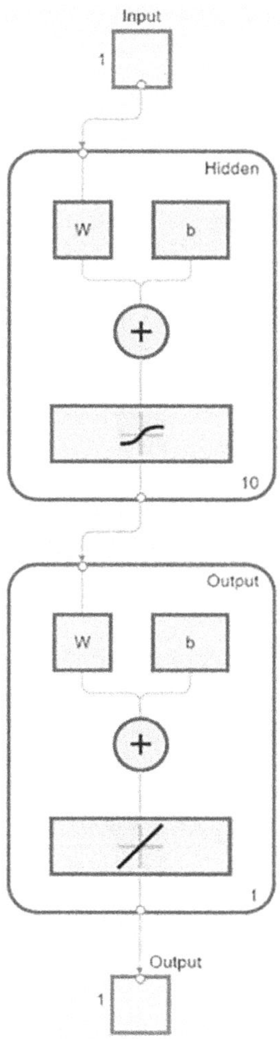

FIGURE 4.4 Architecture of a feedforward neural network for nonlinear function approximation.

data points generated from the function (Equation 4.46), where ϵ is Gaussian noise added to simulate real-world variability. The dashed curve represents the true function $y = 0.5x^3 - 2x^2 + 3x$, providing a reference for the network's expected output.

This dataset serves as the foundation for demonstrating the backpropagation algorithm's capability to learn and approximate complex nonlinear patterns. The close alignment between the data points and the true function highlights the model's learning target and the importance of robust training methodologies.

Figure 4.6 compares the reference (target) samples with the ANN outputs obtained after training. The reference data exhibit a clear nonlinear trend, whereas the ANN output remains nearly constant across the input range, indicating underfitting and insufficient model capacity for the given function approximation task.

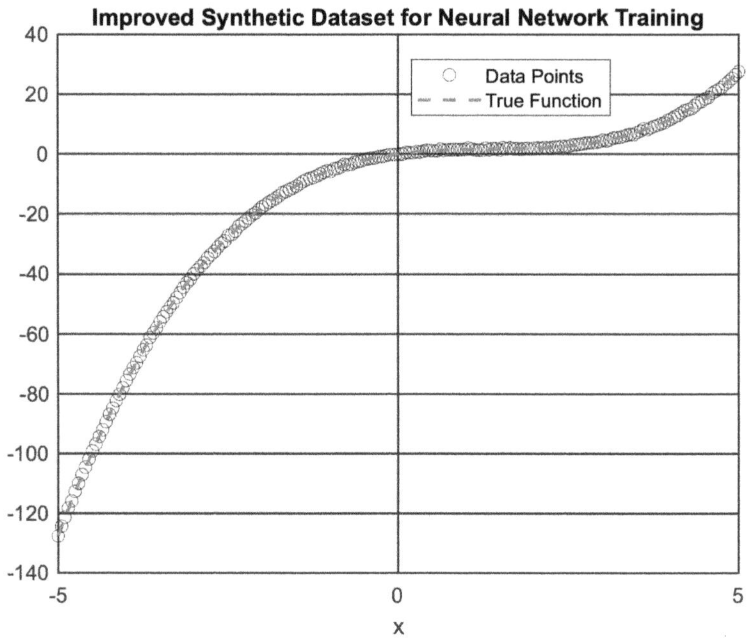

FIGURE 4.5 Synthetic dataset and true function for nonlinear approximation.

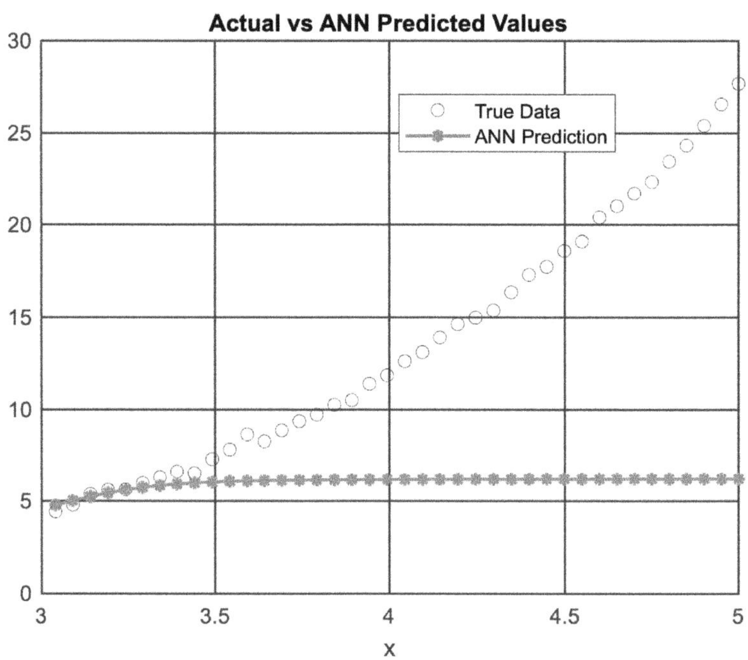

FIGURE 4.6 Comparison of true data and ANN predictions.

This result suggests that the network either:

1. Lacks sufficient complexity (e.g., insufficient neurons or layers) to model the nonlinear function.
2. Requires further tuning of hyperparameters such as the learning rate, number of epochs, or initialization of weights.
3. Might have encountered issues in the training process, such as overfitting or poor convergence.

The figure highlights the importance of properly configuring and training the ANN to ensure it generalizes well to the target function. Adjustments in network architecture or training methodology could significantly improve the model's performance, as discussed in subsequent sections of this chapter.

Figure 4.7 emphasizes the importance of evaluating prediction errors not just globally but also within each data subset to ensure the model is both accurate and robust.

Figure 4.8 compares ANN predictions with target values using a regression plot. The dotted diagonal ($Y=T$) denotes perfect agreement, while the solid regression line represents the best linear fit to the predictions. The large deviation from the diagonal and the nearly flat fitted line indicate that the network mainly produces

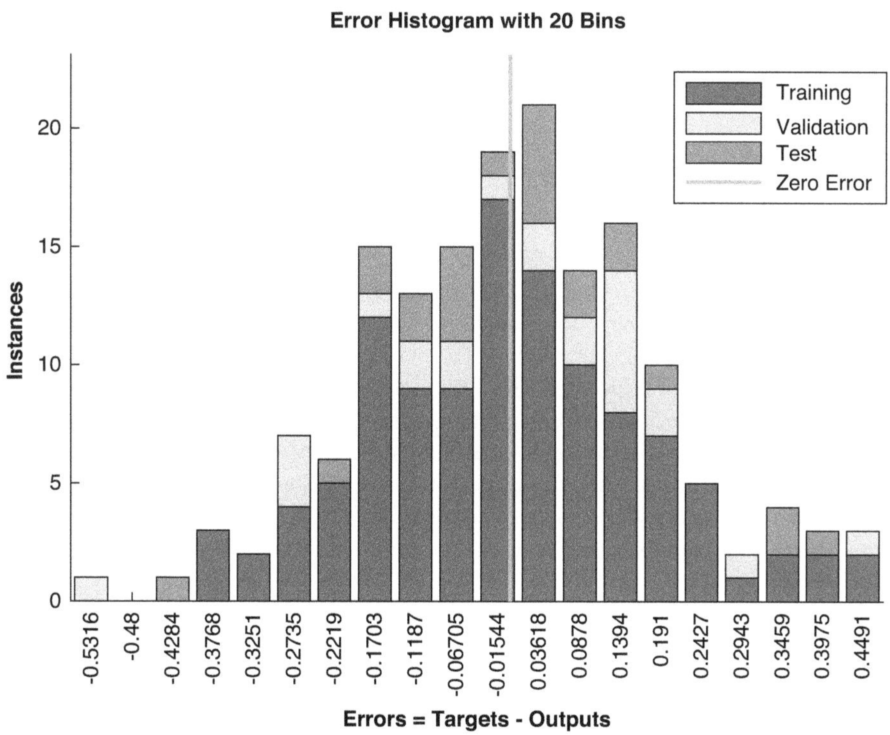

FIGURE 4.7 Error distribution analysis across training, validation, and test phases.

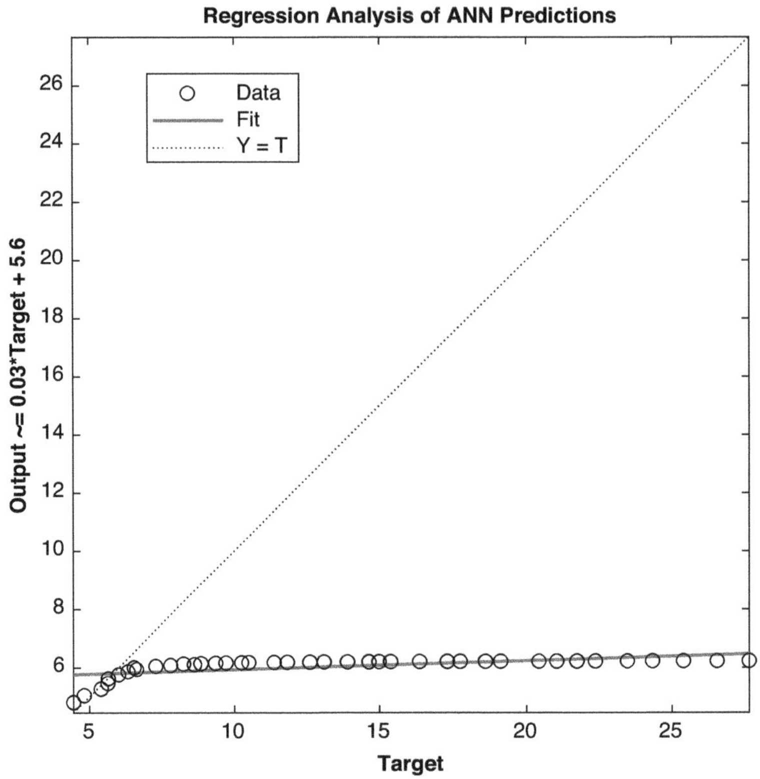

FIGURE 4.8 Regression analysis of neural network performance on target data.

an almost constant output, suggesting underfitting due to limited model capacity or inadequate training.

From the figure, it is evident that the predicted values deviate significantly from the ideal diagonal line, with the linear fit lying flat around a constant value. This result suggests that the neural network failed to capture the underlying relationship in the dataset. The flat regression line and the high intercept highlight the network's inability to generalize beyond a constant prediction, likely due to inadequate training, insufficient complexity in the network architecture, or poor hyperparameter tuning.

Figure 4.9 illustrates the training performance of the neural network measured by the MSE over successive epochs. The three curves correspond to the MSE for the training set, validation set, and test set, showing the progression of error reduction as the model learns.

Key observations from the figure include:

- A sharp decrease in error during the initial epochs, indicating effective learning.
- The model achieves its best validation performance at around 44 epochs, as marked by the circle.

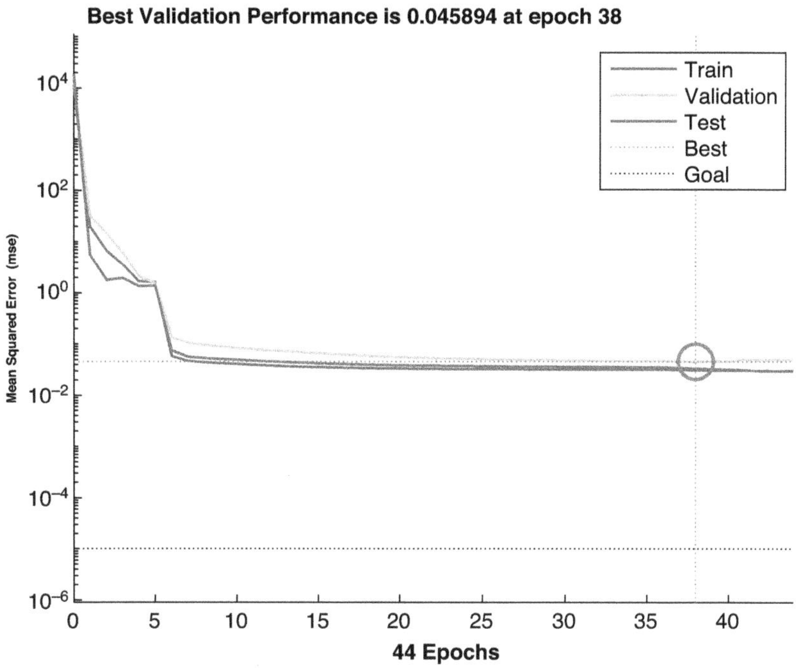

FIGURE 4.9 Training performance of the neural network using MSE.

- The convergence of the training, validation, and test curves suggests the absence of significant overfitting, as the error is consistent across all datasets.

This figure demonstrates the network's ability to minimize errors efficiently while maintaining generalization, as the test set error remains comparable to the training and validation errors.

Figure 4.10 illustrates the progression of key training state variables—gradient, mu, and validation checks—during the neural network optimization process over 44 epochs.

- *Top Panel (Gradient)*: Shows the decrease in gradient magnitude, reflecting how the network adjusts its weights to minimize error. A steady decline indicates convergence toward an optimal solution.
- *Middle Panel (Mu)*: Depicts the variation in the learning parameter *μ*\mu*μ* used in the Levenberg-Marquardt algorithm. Adjustments to μ\muμ help balance the trade-off between convergence speed and stability.
- *Bottom Panel (Validation Fail)*: Tracks validation checks to monitor overfitting. The increase toward the final epochs suggests the model begins to overfit the training data, which triggers an early stopping mechanism.

This figure provides insight into the network's learning dynamics and the mechanisms ensuring effective training while avoiding overfitting.

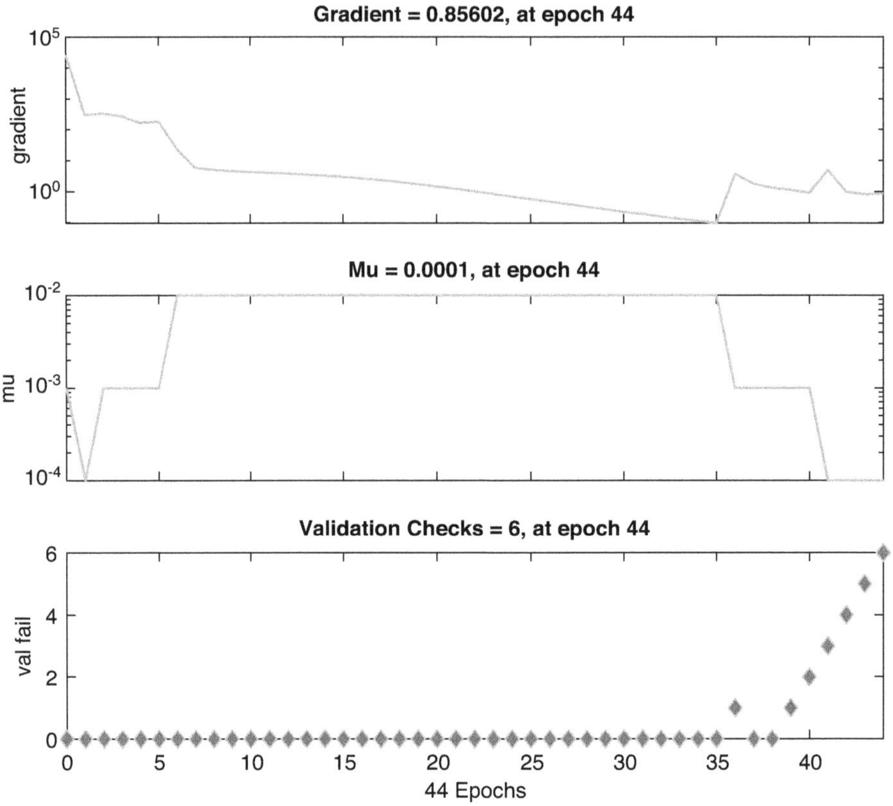

FIGURE 4.10 Training state variables during neural network optimization.

In this section, we provide Python implementations of the analyses and synthetic dataset generation discussed in Chapter 4. These Python scripts are designed to replicate the functionality of the MATLAB® codes, offering an alternative platform for readers familiar with Python. The presented code demonstrates key steps in generating synthetic datasets, training feedforward neural networks, and visualizing results such as error histograms, regression analysis, and performance metrics.

The Python scripts utilize widely used libraries like numpy for numerical computations, matplotlib for visualization, and scikit-learn for constructing and training neural networks. By following these examples, readers will gain practical insights into the process of implementing feedforward neural networks using Python, enabling them to adapt and extend these methods for their specific use cases. These scripts serve as a comprehensive guide, ensuring reproducibility and facilitating further exploration of neural network applications.

```
import numpy as np
import matplotlib.pyplot as plt
from sklearn.model_selection import train_test_split
from sklearn.neural_network import MLPRegressor
```

```
from sklearn.metrics import mean_squared_error, r2_score

# Step 1: Generate Improved Synthetic Dataset
x = np.linspace(-5, 5, 200)  # 200 points between -5 and 5
noise = np.random.randn(len(x)) * 0.2  # Gaussian noise with
higher variance
y = 0.5 * x**3 - 2 * x**2 + 3 * x + noise  # Cubic function
with noise

# Visualize the dataset
plt.figure()
plt.scatter(x, y, color='blue', label='Data Points')
plt.plot(x, 0.5 * x**3 - 2 * x**2 + 3 * x, 'r--',
linewidth=1.5, label='True Function')
plt.grid()
plt.xlabel('x')
plt.ylabel('y')
plt.title('Improved Synthetic Dataset for Neural Network
Training')
plt.legend()
plt.show()

# Step 2: Prepare Data for Neural Network Training
inputs = x.reshape(-1, 1)  # Reshape for proper dimensions
targets = y.reshape(-1, 1)  # Reshape for proper dimensions

# Split data into training (80%) and test (20%) sets
X_train, X_test, Y_train, Y_test = train_test_split(inputs,
targets, test_size=0.2, random_state=42)

# Step 3: Create and Configure the Feedforward Neural
Network
hidden_layer_sizes = (10,)  # Number of neurons in the hidden
layer
model = MLPRegressor(hidden_layer_sizes=hidden_layer_sizes,
max_iter=1000, learning_rate_init=0.01, random_state=42)

# Step 4: Train the Neural Network using Backpropagation
model.fit(X_train, Y_train.ravel())  # Train the model

# Step 5: Test the ANN Model
Y_pred = model.predict(X_test)  # Get predictions from the
trained network
errors = Y_test.ravel() - Y_pred  # Compute prediction errors
performance = mean_squared_error(Y_test, Y_pred)  # Mean
Squared Error (MSE)

# Step 6: Visualize Results

# 6.1 Plot True vs Predicted Values
```

```
plt.figure()
plt.scatter(X_test, Y_test, color='blue', label='True Data')
plt.plot(X_test, Y_pred, 'r-*', linewidth=1.5, label='ANN
Prediction')
plt.grid()
plt.xlabel('x')
plt.ylabel('y')
plt.title('Actual vs ANN Predicted Values')
plt.legend()
plt.show()

# 6.2 Error Histogram
plt.figure()
plt.hist(errors, bins=20, color='blue', edgecolor='black')
plt.title('Error Histogram')
plt.xlabel('Prediction Error')
plt.ylabel('Frequency')
plt.grid()
plt.show()

# 6.3 Regression Plot
plt.figure()
plt.scatter(Y_test, Y_pred, color='blue', label='Data')
plt.plot(Y_test, Y_test, 'k--', label='Ideal Fit')
plt.grid()
plt.xlabel('True Values')
plt.ylabel('Predicted Values')
plt.title('Regression Analysis of ANN Predictions')
plt.legend()
plt.show()

# Display training performance
print(f'Training Performance (MSE): {performance:.5f}')
```

5 Dynamical Neural Networks

5.1 RECURRENT NEURAL NETWORKS

Recurrent Neural Networks (RNNs) are neural networks designed for sequential or time-dependent data. At each time step, an RNN combines the current input with an internal state that carries information from previous steps, allowing the model to capture temporal dependencies. For comparison with classical time-series forecasting, we also refer to the Autoregressive Integrated Moving Average (ARIMA) model, which combines autoregressive terms, differencing, and moving-average components. A widely used RNN variant is the Long Short-Term Memory (LSTM) network, which utilizes gated memory units to better retain long-range information and mitigate vanishing gradient effects. Depending on where the feedback is applied, RNNs are commonly categorized as:

- *Jordan-Type RNN*: This type of RNN has feedback connections from the output layer to the context layer. The feedback connections in Jordan-type RNNs can be particularly useful for specific applications.
- *Elman-Type RNN*: Elman-type RNNs incorporate feedback connections from hidden layers to the context layer. These feedback connections enable the network to use previously computed hidden states.

In both Jordan and Elman-type RNNs, the Back Propagation Algorithm is commonly employed as the learning algorithm. The dynamic properties of these neural network topologies are integrated into the Back Propagation Algorithm, making them suitable for a wide range of tasks that require sequential or time-dependent information processing.

5.1.1 Jordan-Type RNN

In a Jordan-type Recurrent Neural Network, the feedback connections originate from the output layer and are directed back to the input side through a specialized set of nodes known as the context layer. This layer is structurally distinct from the standard input nodes and plays a unique role in retaining information from previous outputs. By integrating this feedback mechanism, the network is able to incorporate temporal dependencies into its current processing.

Figure 5.1 illustrates the architecture of the Jordan-type RNN, emphasizing the feedback pathway from the output layer to the context layer. Serving as a form of short-term memory, the context layer allows the network to utilize past information when generating new outputs, making it especially useful in time-dependent learning tasks.

DOI: 10.1201/9781003673972-5

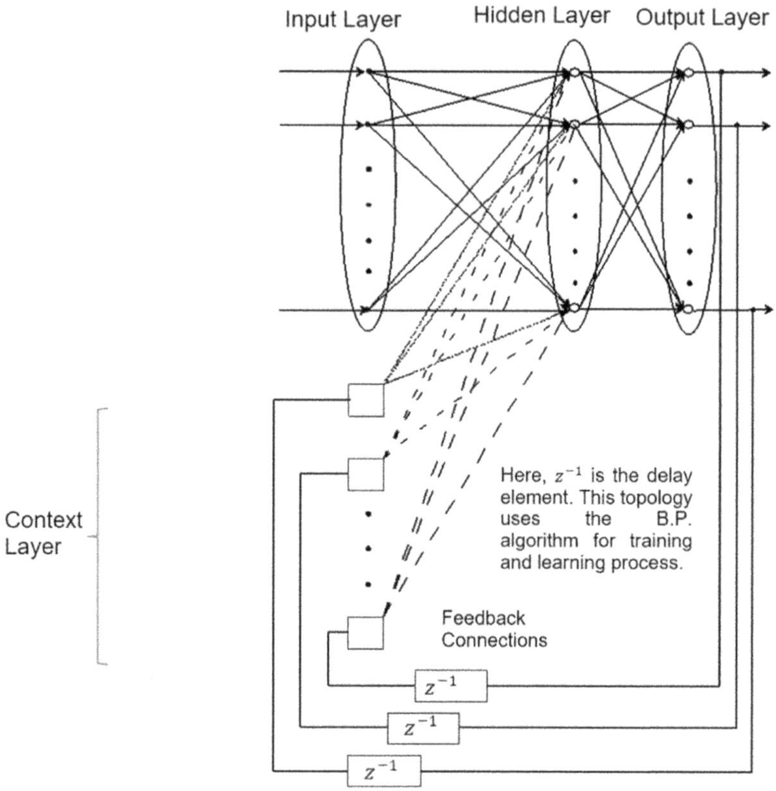

FIGURE 5.1 Jordan's Recurrent Neural Net.

5.1.2 Elman-Type RNN

In this case, the feedback connections are provided from the hidden layer(s). Consider a simple case having only one hidden layer as given below:

The Elman-type Recurrent Neural Network introduces a feedback structure that significantly enhances the network's ability to model dynamic behavior over time. Unlike the Jordan-type RNN, Elman networks utilize feedback connections from the hidden layer to a dedicated context layer, enabling the reuse of internal features across multiple time steps. This architecture is particularly well-suited for applications involving temporal dependencies, such as system diagnostics, time-series prediction, and speech recognition. Its structure supports deeper learning from past internal states, allowing the network to detect recurring patterns, adapt to evolving input conditions, and improve performance in tasks where the memory of previous events is essential.

Figure 5.2 illustrates the structure of an Elman-type RNN, where feedback is provided from the hidden layer to a context layer. This architecture enhances the network's ability to reuse internal features, allowing for the repeated evaluation of extracted information. As discussed in the following paragraph, Elman networks

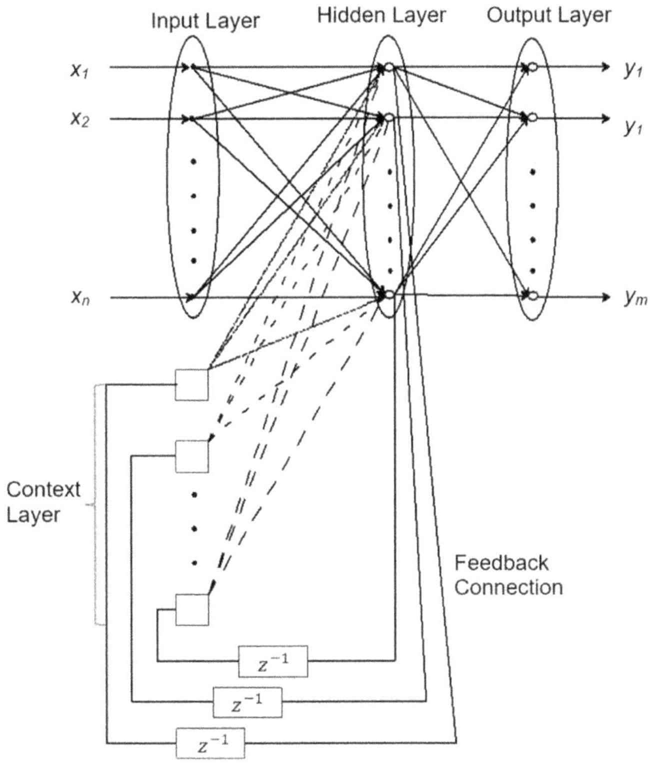

FIGURE 5.2 Elman's Recurrent Neural Net.

offer an advantage over Jordan networks by leveraging hidden node feedback, which makes them particularly effective in fault detection tasks. Their dynamic structure contributes significantly to recognizing recurring patterns in sequential data.

In this context, these topologies also utilize the Back Propagation Algorithm for learning and training procedures. In other words, these topologies, or Recurrent Neural Networks, are commonly referred to as Dynamical Neural Networks.

As a point of comparison, the Elman type of RNN has an advantage over the Jordan type because it incorporates feedback connections from hidden nodes, allowing it to make use of extracted information multiple times. These RNN architectures have been found to be particularly effective in fault detection problems. In general, they play a crucial role in extracting valuable information or similar patterns from data.

5.2 APPLICATION: TIME-SERIES PREDICTION USING RECURRENT NEURAL NETWORKS

Recurrent Neural Networks are powerful tools for modeling sequential data and are widely used in applications such as weather forecasting, stock market analysis, and energy system monitoring. In this section, we apply RNNs to a synthetic time-series dataset to demonstrate their ability to capture temporal patterns and dependencies effectively.

The synthetic data used in this application is generated using a mathematical function: $y(t) = \sin(t) + 0.5 \cdot \cos(2t)$, where t represents time in seconds. This equation blends sinusoidal and cosine components, creating a periodic sequence that mimics the variability and complexity of real-world time-series data. By introducing multiple frequency components, the dataset provides a controlled and structured testbed for analyzing RNN behavior.

In this application, the dataset is divided into input–output pairs. Each input consists of a sequence of observed values, and the corresponding output is the subsequent value in the sequence. The goal is to train an RNN to model the underlying temporal relationships and predict future values based on past observations. The network architecture includes a recurrent layer with feedback loops to retain memory of previous time steps and a dense output layer for generating precise predictions.

The analysis involves training the RNN on 80% of the dataset and evaluating its performance on the remaining 20%. Key metrics, such as Root Mean Squared Error (RMSE), are used to assess the model's predictive accuracy. The results, including visualizations of actual vs. predicted values and error distributions, highlight the strengths and limitations of the RNN in handling time-series prediction tasks.

This example provides a hands-on demonstration of how RNNs can be effectively implemented for time-series prediction, offering readers a practical understanding of their capabilities and laying the groundwork for exploring real-world applications.

Figure 5.3 illustrates the synthetic time-series data generated using a combination of sinusoidal and cosine functions, representing periodic patterns over time. The dataset is constructed using the equation $y(t) = \sin(t) + 0.5 \cdot \cos(2t)$, where t is the time in seconds. This mathematical representation creates a smoothly varying waveform

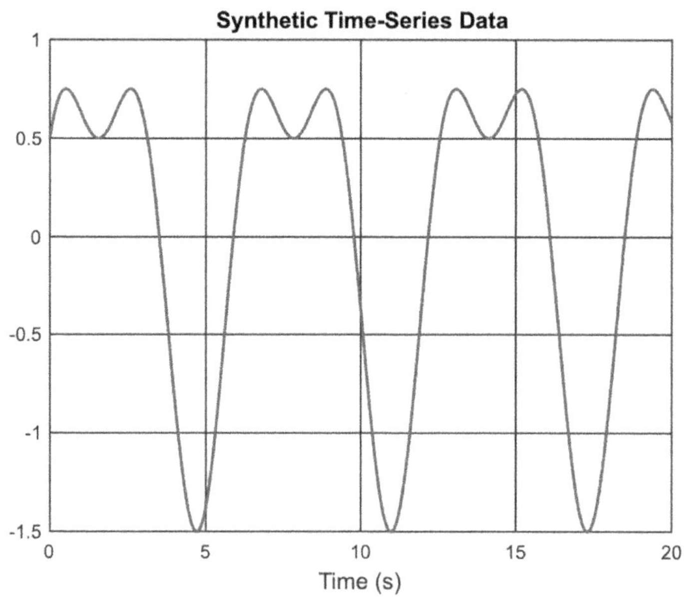

FIGURE 5.3 Synthetic time-series data generated using sinusoidal and cosine functions.

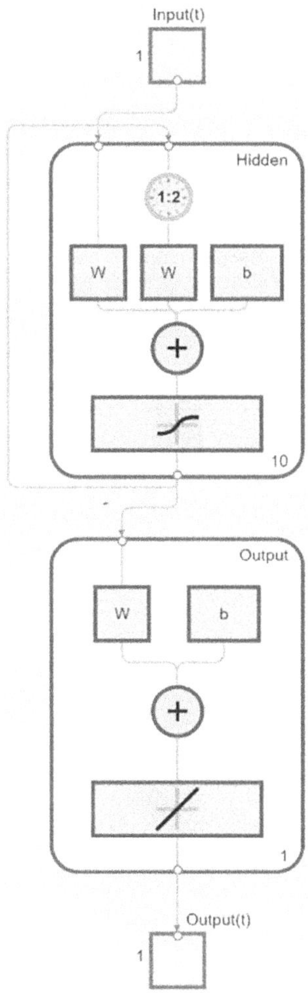

FIGURE 5.4 Recurrent Neural Network architecture for time-series prediction.

that mimics real-world temporal dynamics, making it suitable for testing Recurrent Neural Networks in time-series prediction tasks. The visualization highlights the variability and periodicity embedded in the synthetic dataset, which serves as the input for further RNN-based analyses in this chapter.

Figure 5.4 illustrates the architecture of the Recurrent Neural Network used for time-series prediction in this application. The model, specifically an *Elman RNN*, incorporates feedback loops within the hidden layer, enabling it to capture temporal dependencies in sequential data. The network consists of a hidden layer with 10 units, each equipped with weights (*W*) and biases (*b*), followed by an output layer that predicts the next value in the time series. This structure is particularly suited for modeling dynamic patterns in data where past observations influence future predictions.

Table 5.1 provides an overview of the RNN model's training progress, highlighting key metrics such as the number of epochs, elapsed time, performance, and gradient

TABLE 5.1
Training Progress Summary for the RNN Model

Metric	Initial Value	Stopped Value	Target Value
Epoch	0	100	100
Elapsed time (s)	0.01	210.64	N/A
Performance	6.13	0.0146	0
Gradient	0.001	1e−05	N/A
Validation checks	0	5	5

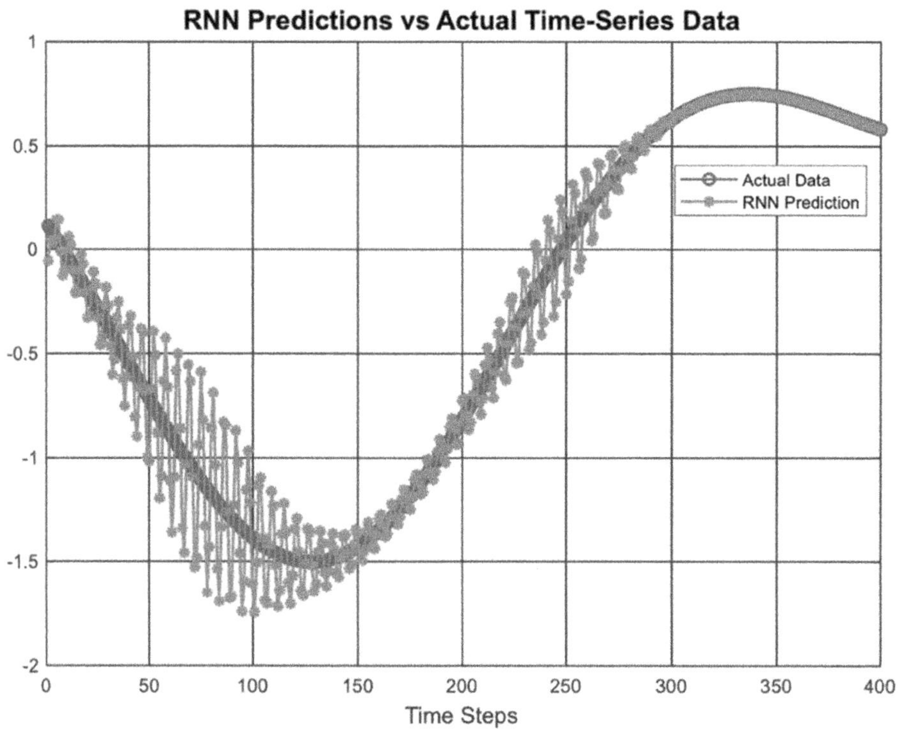

FIGURE 5.5 Comparison of actual data and RNN predictions.

descent values. The model was trained for 100 epochs, achieving a performance value close to the target threshold. This summary demonstrates the effectiveness of the training process, including the optimization of weights and biases to minimize prediction errors. The table serves as a concise record of the model's convergence and readiness for evaluation.

Figure 5.5 shows how the Recurrent Neural Network (RNN) performs when predicting sequential data generated from synthetic sine and cosine functions. The actual time-series values are plotted together with the RNN's predicted output. The close agreement between the two curves indicates that the model has learned the

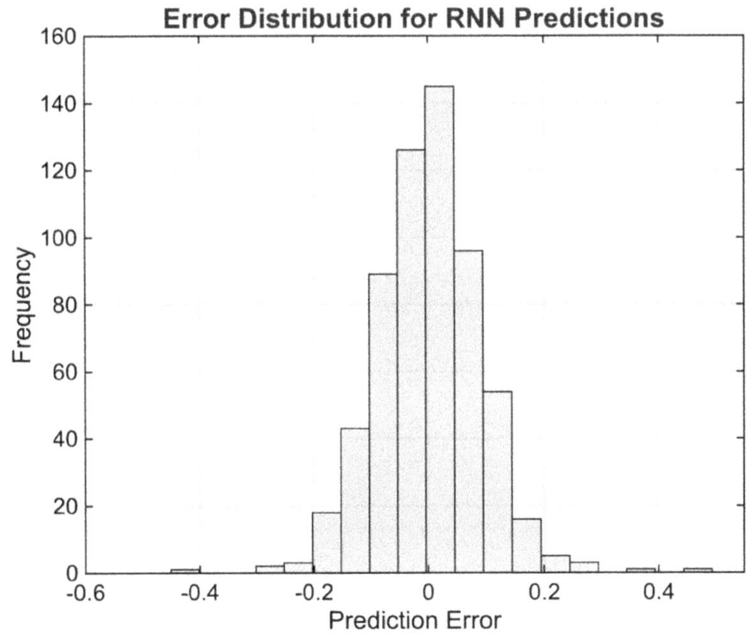

FIGURE 5.6 Prediction error distribution for the RNN model.

underlying temporal structure of the signal and can reproduce it with good accuracy. Overall, the figure suggests that the trained RNN can forecast future samples reliably when the data exhibit smooth, periodic patterns.

Figure 5.6 presents the distribution of prediction errors for the Recurrent Neural Network model applied to the synthetic time-series dataset. The histogram shows that the majority of prediction errors are concentrated around zero, indicating high accuracy and minimal deviation between the actual and predicted values. This tightly clustered error distribution highlights the model's effectiveness in capturing the temporal patterns of the data, further validating the robustness of the RNN for time-series analysis tasks.

Figure 5.7 illustrates the progression of the Mean Squared Error during the training phase of the Recurrent Neural Network model over six epochs. The rapid and consistent decline in Mean Squared Error highlights the model's ability to learn effectively from the synthetic time-series data. The convergence of the error toward the predefined performance goal demonstrates the efficiency of the training algorithm and the suitability of the model architecture for the task. The visualization provides a clear indication of the training process's success in minimizing prediction errors.

Figure 5.8 shows the distribution of training errors obtained from the Recurrent Neural Network (RNN). The histogram summarizes how far the model's predictions deviate from the true values across the training samples. A vertical reference line marks the zero-error point, which corresponds to a perfect match between prediction

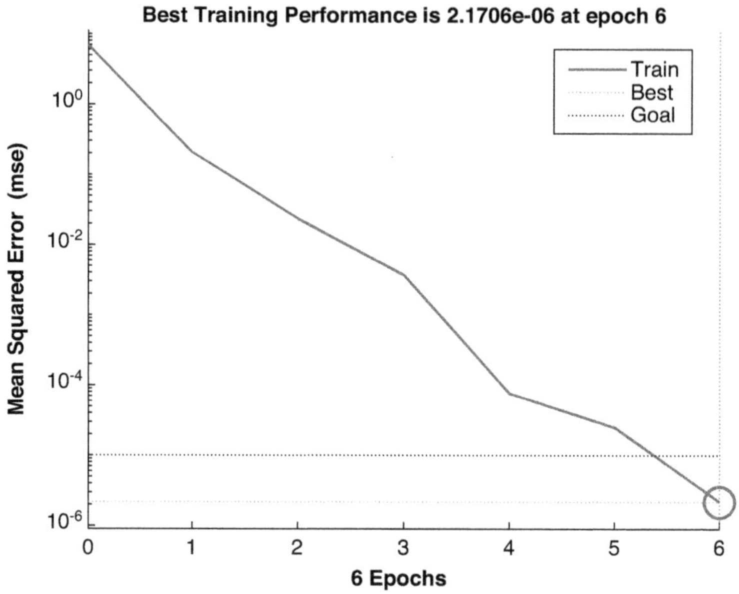

FIGURE 5.7 Mean squared error progression during RNN training.

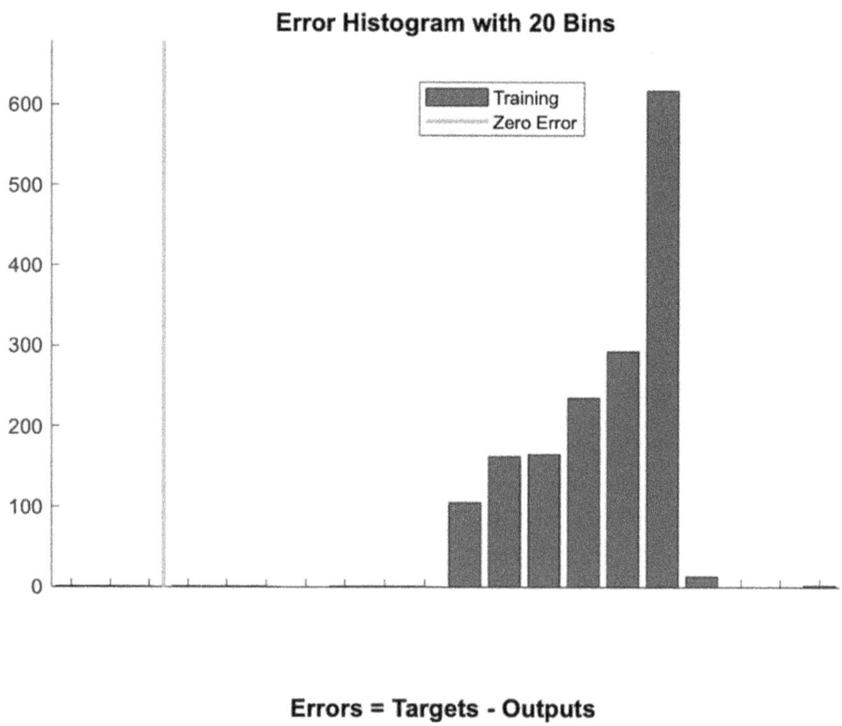

FIGURE 5.8 Training error distribution of RNN predictions.

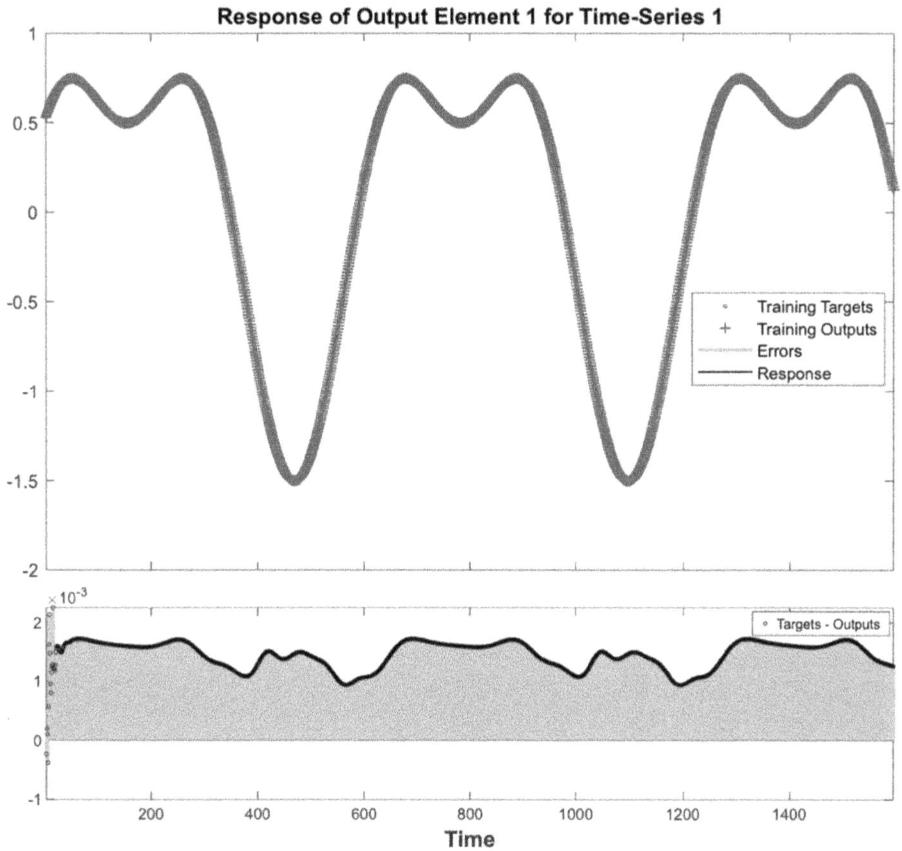

FIGURE 5.9 Comparison of training outputs and targets with error distribution.

and target. The strong clustering of errors around this reference indicates that most predictions have small deviations, suggesting effective learning during training. Overall, the figure supports that the model captures the main structure of the synthetic time-series signal with stable performance.

Figure 5.9 shows the training performance of the Recurrent Neural Network. The upper plot compares the network outputs, shown with '+' markers, against the training targets, shown with '.' markers. The close agreement between these markers indicates that the model learns the time-dependent structure of the synthetic sequence and tracks the target response closely.

The lower plot presents the prediction error at each time step, computed as the difference between the targets and the network outputs. The errors stay small across the sequence, indicating stable learning and a good fit to the training data for this time-series prediction task.

The analyses presented in this section were performed using a custom Python script that implements the Recurrent Neural Network for time-series prediction. The script includes synthetic data generation, model training, and performance evaluation through metrics and visualizations. Each step, from preparing the dataset to validating the model's predictive accuracy, has been meticulously designed to illustrate the application of RNNs in handling temporal patterns. The Python code for reproducing these analyses is provided below, ensuring reproducibility and enabling readers to experiment further.

```
import numpy as np
import matplotlib.pyplot as plt
from sklearn.model_selection import train_test_split
from tensorflow.keras.models import Sequential
from tensorflow.keras.layers import SimpleRNN, Dense

# 1. Generate Synthetic Time-Series Data
t = np.arange(0, 20, 0.01)  # Time vector (0 to 20 seconds
with 0.01-second intervals)
synthetic_data = np.sin(t) + 0.5 * np.cos(2 * t)  # Synthetic
data (sinusoidal combination)
input_sequence = synthetic_data[:-1]  # Inputs (all except the
last point)
target_sequence = synthetic_data[1:]  # Targets (all except
the first point)

# Visualize the synthetic time-series data
plt.figure(figsize=(8, 4))
plt.plot(t, synthetic_data, linestyle='-', linewidth=1.5)
plt.grid(True)
plt.xlabel('Time (s)', fontsize=12)
plt.ylabel('Value', fontsize=12)
plt.title('Synthetic Time-Series Data', fontsize=14)
plt.show()

# 2. Prepare Data for RNN
# Split data into training (80%) and testing (20%) sets
X_train, X_test, Y_train, Y_test = train_test_split(
    input_sequence, target_sequence, test_size=0.2,
random_state=42
)

# Reshape data for RNN (samples, timesteps, features)
X_train = X_train.reshape(-1, 1, 1)  # Input shape: (samples,
timesteps, features)
X_test = X_test.reshape(-1, 1, 1)
Y_train = Y_train.reshape(-1, 1)
Y_test = Y_test.reshape(-1, 1)

# 3. Build the RNN Model
model = Sequential()
model.add(SimpleRNN(10, activation='tanh', input_shape=(1, 1),
return_sequences=False))  # RNN layer with 10 units
model.add(Dense(1))  # Output layer with 1 neuron
model.compile(optimizer='adam', loss='mean_squared_error')  #
Compile the model
```

```
# 4. Train the RNN
history = model.fit(
    X_train, Y_train,
    epochs=100,
    batch_size=32,
    validation_data=(X_test, Y_test),
    verbose=1
)

# 5. Test the RNN
Y_pred = model.predict(X_test)

# 6. Visualize RNN Predictions vs Actual Data
plt.figure(figsize=(8, 4))
plt.plot(Y_test, linestyle='-', marker='o', markersize=4,
linewidth=1.2, label='Actual Data')
plt.plot(Y_pred, linestyle='--', marker='*', markersize=6,
linewidth=1.2, label='RNN Prediction')
plt.grid(True)
plt.xlabel('Time Steps', fontsize=12)
plt.ylabel('Value', fontsize=12)
plt.title('RNN Predictions vs Actual Time-Series Data',
fontsize=14)
plt.legend()
plt.show()

# 7. Error Analysis
rmse = np.sqrt(np.mean((Y_test - Y_pred) ** 2))
print(f"Root Mean Squared Error (RMSE): {rmse:.4f}")

# 8. Plot Error Histogram
errors = (Y_test - Y_pred).flatten()
plt.figure(figsize=(8, 4))
plt.hist(errors, bins=20, histtype='step', linewidth=1.5)
plt.grid(True)
plt.xlabel('Prediction Error', fontsize=12)
plt.ylabel('Frequency', fontsize=12)
plt.title('Error Distribution for RNN Predictions',
fontsize=14)
plt.show()
```

5.3 COMPARATIVE APPLICATION: ARMA VS. RNN FOR TIME-SERIES PREDICTION

In this section, we aim to compare the predictive capabilities of Recurrent Neural Networks with a well-established statistical approach: the AutoRegressive Moving Average (ARMA) model. While RNNs are designed to capture complex temporal dependencies through iterative learning, ARMA models offer a more traditional strategy, relying on past values and error terms to forecast future observations.

The ARMA model consists of two main components:

- The AutoRegressive (AR) part models the current value of the time series as a linear combination of its previous values.
- The Moving Average (MA) part corrects predictions using past forecasting errors.

By applying the ARMA model to the same synthetic dataset used in Section 5.2, we establish a meaningful benchmark for evaluating both traditional and data-driven approaches. The analysis explores the strengths and weaknesses of each method, highlighting their suitability under different data conditions.

Figure 5.10 illustrates how ARMA and RNN models follow the ground truth time series. While RNN appears to capture the general shape and trend more flexibly, ARMA offers smoother, more conservative forecasts. The figure helps to visually compare the models' performances across the full test set.

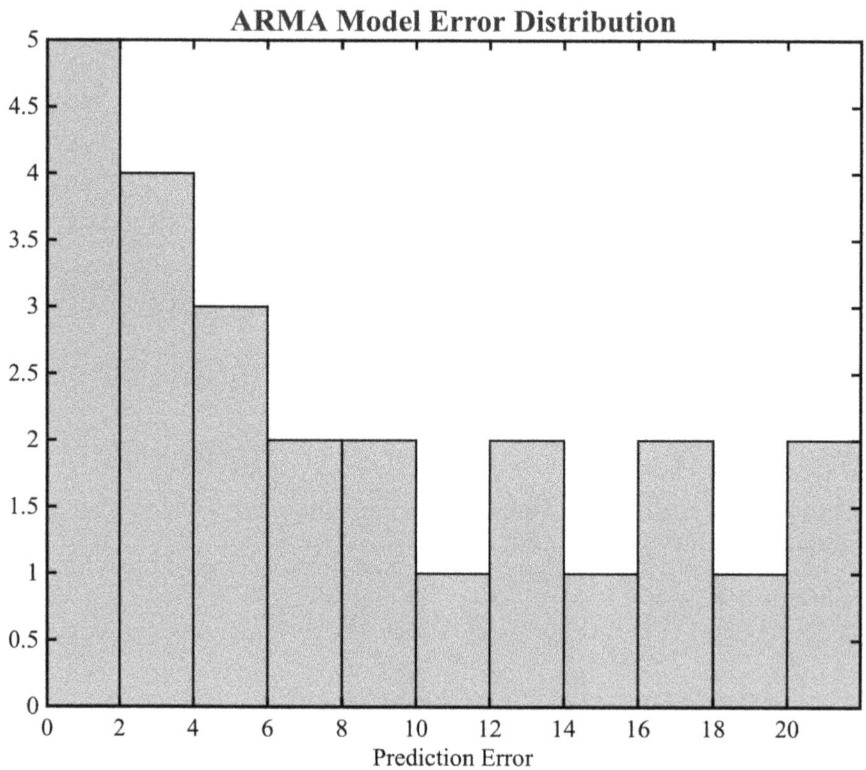

FIGURE 5.10 Prediction error distribution for the ARMA model.

TABLE 5.2
ARIMA(2,0,2) Model Results (Gaussian Distribution)

Parameter	Value	Standard Error	*T*-Statistic	*P*-Value
Constant	0.10219	0.016232	6.2958	3.0577e−10
AR{1}	1.4958	0.011995	124.7	0
AR{2}	−0.50805	0.011826	−42.961	0
MA{1}	−0.84225	0.060404	−13.944	3.4442e−44
MA{2}	0.32443	0.065884	4.9243	8.4657e−07
Variance	0.092326	0.010154	9.0928	9.6546e−20

RMSE for the ARMA Model: 9.77380.

TABLE 5.3
Comparative Performance of ANN and ARMA Models

Model	RMSE	Strengths	Limitations
ANN	4.2531	Captures nonlinear relationships, robust for complex data	Requires large data, computationally intensive
ARIMA(2,0,2) (ARMA Model)	9.7738	Good for stationary time-series, interpretable	Struggles with nonlinear patterns, less flexible

Table 5.2 presents the parameter estimates for the ARIMA(2,0,2) model applied to the synthetic dataset. The results indicate that the AR and MA terms significantly contribute to capturing the time-dependent structure of the data. The AR terms suggest a strong dependency on past values, with AR{1} exhibiting a dominant effect, as indicated by its large magnitude and statistically significant t-statistic. Similarly, the MA terms demonstrate the ability to correct for noise and residuals, with MA{1} playing a more substantial role in the model's performance.

The RMSE for the ARMA model is 9.7738, reflecting the model's predictive accuracy. This value will be compared with the RMSE of the RNN to evaluate the effectiveness of both approaches. The low variance value suggests that the model's residuals exhibit minimal fluctuations, indicating a well-fitted model. The statistical significance of the estimated parameters (as denoted by their p-values) further validates the robustness of the model in capturing the temporal dependencies present in the dataset. This comparison provides valuable insights into the strengths and limitations of traditional statistical methods versus deep learning approaches in time-series forecasting.

As shown in Table 5.3, Artificial Neural Network (ANN) models outperform ARMA in RMSE, showcasing their capability to learn complex, nonlinear patterns. However, this comes at the cost of data hunger and higher computational overhead. In contrast, ARMA remains attractive for applications prioritizing simplicity and interpretability.

5.4 MODEL SELECTION BASED ON DATA CHARACTERISTICS

Choosing the appropriate time-series forecasting method is not solely a matter of performance metrics such as RMSE; it also requires a clear understanding of the data's underlying structure. Characteristics such as linearity, stationarity, continuity, and sequence-dependency play a critical role in guiding model selection.

5.4.1 Linearity vs. Nonlinearity

- Linear time series exhibit patterns that can be effectively modeled using statistical techniques like AR, MA, ARMA, or ARIMA. These models assume that the relationship between past and present values is linear and can be captured through AR and MA terms.
- Nonlinear data, on the other hand, often contain complex dependencies, seasonal effects, or structural breaks that linear models fail to capture. In such cases, machine learning models—particularly ANN and RNN—are better suited as they learn patterns from data without requiring strong parametric assumptions.

5.4.2 Continuity vs. Discreteness

- Continuous time series (e.g., temperature, energy consumption) are typically smooth and can be processed by most forecasting models, including statistical and neural approaches.
- Discrete or categorical time series (e.g., event logs, state transitions, binary outcomes) often require sequential memory and classification capabilities, making RNNs, LSTMs, or even Support Vector Machines more appropriate choices (Tables 5.4 and 5.5).

TABLE 5.4
Comparison of Time-Series Forecasting Methods

Method	Description	Strengths	Limitations	Typical Applications	RMSE (This Study)
AR	Predicts current values based on a linear combination of previous values.	Simple, interpretable, effective on stationary data.	Cannot model noise or external shocks.	Short-term prediction of stable trends.	e.g., 12.5

(Continued)

TABLE 5.4 (*Continued*)
Comparison of Time-Series Forecasting Methods

Method	Description	Strengths	Limitations	Typical Applications	RMSE (This Study)
MA	Predicts based on past error terms.	Smooths short-term fluctuations.	Does not capture long-term structure.	Noise reduction in signals.	e.g., 11.3
ARMA	Combines AR and MA for modeling linear temporal structure.	Balances trend and noise modeling.	Not suitable for non-stationary data.	Intermediate-level time series.	9.77
ARIMA	Adds differencing to ARMA for non-stationary series.	Handles trending data effectively.	Poor performance on nonlinear patterns.	Forecasting in economics, energy, environment.	8.45
ANN	Learns complex nonlinear patterns using layered networks.	Flexible, captures hidden relationships.	Requires large data and high computation.	Multivariate, nonlinear series forecasting.	4.25
RNN	Incorporates memory of previous inputs for sequential modeling.	Suitable for long-term dependencies.	Complex training; vanishing gradient risk.	Speech, signal, and multistep time-series tasks.	3.89

TABLE 5.5
Model Recommendation Matrix Based on Data Type

Data Type	Best Suited Methods	Reasoning
Linear + stationary	AR, MA, ARMA	Captures simple lagged relationships.
Linear + non-stationary	ARIMA	Handles trends via differencing.
Nonlinear + continuous	ANN, RNN	Learns complex dynamic behaviors.
Nonlinear + discrete	RNN, LSTM, Support Vector Machines (if labeled)	Suited for classification and sequence modeling.

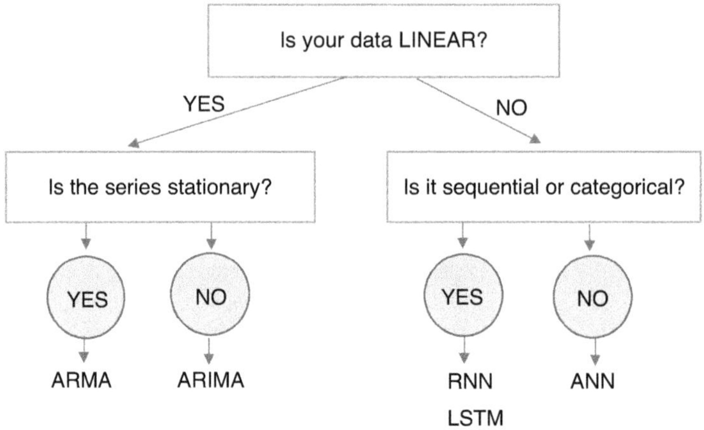

FIGURE 5.11 Model selection decision tree based on time-series data characteristics.

5.4.3 Model Selection Flow (Conceptual Guide)

Below is a simplified decision guide that helps match the forecasting method to the nature of the data (Figure 5.11).

This decision tree highlights that while classical models are ideal for structured and predictable systems, data exhibiting nonlinearity or complex temporal dependencies benefits greatly from deep learning models that can extract hidden patterns.

6 System Identification and Use of Neural Networks

6.1 SYSTEM IDENTIFICATION (MODELING)

A system can be described by its order and parameters. This is observed in the mathematical modeling of the system. Depending on the system type, such as continuous or discrete. This mathematical model is represented by a differential equation for continuous and discrete systems, respectively. Now, consider a general linear model as a discrete system (Figure 6.1):

$$\begin{aligned} a_0 y[n] + a_1 y[n-1] + a_1 y[n-1] + \cdots + a_N y[n-N] \\ = b_0 x[n] + ab_1 x[n-1] + b_1 x[n-1] + \cdots + b_M x[n-M] \end{aligned} \tag{6.1}$$

where

$$\sum_{k=0}^{N} a_k y[n-k] = \sum_{m=0}^{M} b_m x[n-m] \tag{6.2}$$

This is the linear combination of input and output pairs.

Here, $a_0 \triangleq 1$. If so, it is written as

$$y[n] = -\sum_{k=1}^{N} a_k y[n-k] + \sum_{m=0}^{M} b_m x[n-m] \tag{6.3}$$

where the main problem is the determination of the system parameters, and also the determination of the correct system order N.

6.1.1 For the Stochastic System

The following block diagram can represent the stochastic system model (Figure 6.2).

Here,

ε_t: Independent random error

$\{\varepsilon_t\}$: The set of random processes, such as white noise.

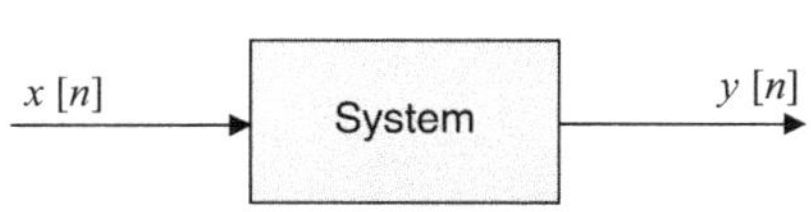

FIGURE 6.1 Discrete-time SISO system.

DOI: 10.1201/9781003673972-6

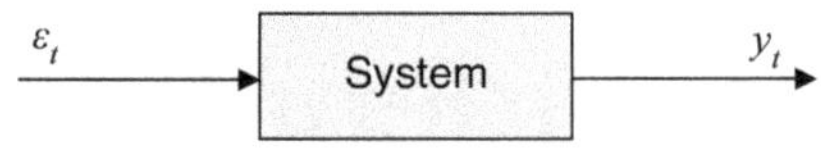

FIGURE 6.2 A linear stochastic system.

Statistically $\{\varepsilon_t\} \sim \text{Dist}\left(0,\sigma^2\right)$.

Therefore, observing the system outputs (current and former values of the system outputs), the modeling task can be used to characterize the process:

$$\varepsilon_t = \left|y_t - y_t^*\right| \tag{6.4}$$

where

y_t^*: Model output
y_t: Actual output

For the auto-regressive process, it becomes

$$y_t + a_1 y_{t-1} + a_2 y_{t-2} + \cdots + a_p y_{t-N} = \varepsilon_t \tag{6.5}$$

Here, ε_t can be interpreted as a white noise process.

For the determination of the system parameters $a_1, a_2, \ldots, a_N$, some special techniques/methods are used to determine the system parameters. In this manner, one is the well-known Yule-Walker method (equation). Here $a_i : i = 1, \ldots, N$ are called the auto-regressive parameters at order N. The Yule-Walker equation is given as

$$\left[\underbrace{\begin{matrix} R_0 & R_1 & \cdots & R_{N-1} \\ R_1 & R_0 & \cdots & R_{N-2}, \\ \vdots & \vdots & \ddots & \vdots \\ R_{N-1} & R_{N-2} & \cdots & R_0 \end{matrix}}_{\text{Auto correlation Matrix}}\right]\left[\underbrace{\begin{matrix} a_1 \\ a_2 \\ \vdots \\ a_N \end{matrix}}_{\text{Coefficients}}\right] = \begin{bmatrix} R_1 \\ R_2 \\ \vdots \\ R_N \end{bmatrix} \tag{6.6}$$

Thus, each data sample can be predicted as

$$\hat{y}_t = -\sum_{i=1}^{N} \hat{a}_i y_{t-i} \tag{6.7}$$

Then the residue or difference between the actual and model outputs is $\hat{\varepsilon}_t$ and it becomes

$$\hat{\varepsilon}_t = y_t - \hat{y}_t \tag{6.8}$$

For this computation, the system order p can be calculated using Akaike's Information Criteria.[1] In this computation, the matrix elements R_τ are given as

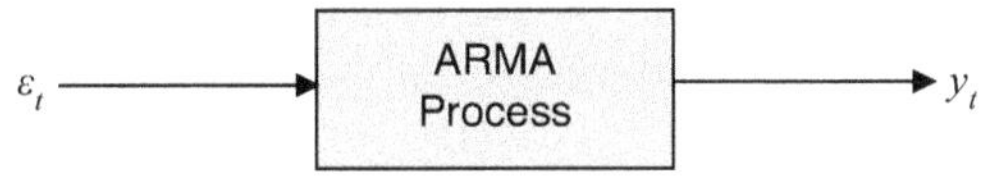

FIGURE 6.3 ARMA modeling.

$$R_\tau = \frac{1}{N}\sum_{t=\tau+1}^{N} y_t y_{t-\tau} \tag{6.9}$$

Finally, if it is determined that the system order and system parameters are known, the approach is known as 'Identification.' Therefore, the system is defined in order and with all parameters.

6.1.2 ARMA Process (Model)

The ARMA model can also be written in the form of Equation (6.1) using the random input like Equation (6.10). Hence, it becomes (Figure 6.3)

$$y_t - a_1 y_{t-1} - \ldots - a\_N\ y_t\_N = \varepsilon_t + b_1\varepsilon_{t-1} + \ldots + b\_M\ \varepsilon_t\_M. \tag{6.10}$$

where ε_t can be shown as a white noise process:

$$\varepsilon_t \sim W\left(0,\sigma^2\right) \tag{6.11}$$

W is a white noise process with zero mean value and variance σ^2.
or equivalently,

$$y_t - \sum_{i=1}^{N} a_i y_{t-i} = \varepsilon_t + \sum_{i=1}^{M} b_i \varepsilon_{t-i} \tag{6.12}$$

Here,

N: autoregressive (AR) order
M: moving-average (MA) order
y_t: output (time-series value) at time t
ε_t: white-noise error term at time t
a_1: AR coefficients
b_1: MA coefficients
$W(0, \sigma^2)$: zero-mean white-noise process with variance σ^2

6.2 USE OF NNs IN SYSTEM MODELING

Here, NNs are feedforward networks that use the Back Propagation algorithm. A simple NN model is in the form of an observer (Figure 6.4).

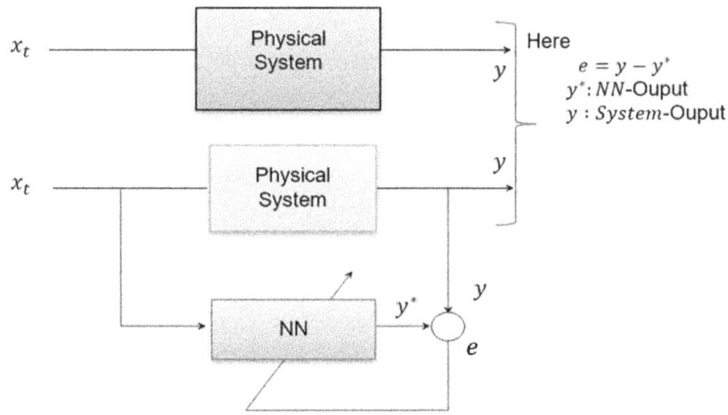

FIGURE 6.4 Physical System and NN-based observer.

If the modeling error reaches an acceptable value at the end of the learning procedure, the NN topology will emulate the actual physical system. For this reason, the system behaves as an observer.

6.3 APPLICATIONS OF ADAPTIVE NNs IN SYSTEM IDENTIFICATION

An ANN is an adaptive system. Therefore, it can be defined in the Adaptive NN Systems configuration. Adaptivity of NNs can be used in the form of

1. System Identification and Modeling
2. Inverse Modeling
3. Adaptive Interference Canceling
4. Adaptive Prediction and noise canceling

6.3.1 System Identification and Modeling

The NN model uses a parallel connection to the system. A simple modeling technique is shown in the following block diagram as an observer (Figure 6.5).

The modeling system with delay z^{-1} and noise is given as below (Figure 6.6).

Here,

e: Difference or error

6.3.2 Inverse Modeling

Inverse modeling uses a series of connections with the system. A simple inverse modeling system is given as (Figure 6.7).

Here, the error term is the difference between the x and x^*, and then, the NN topology is formed with the help of this error term.

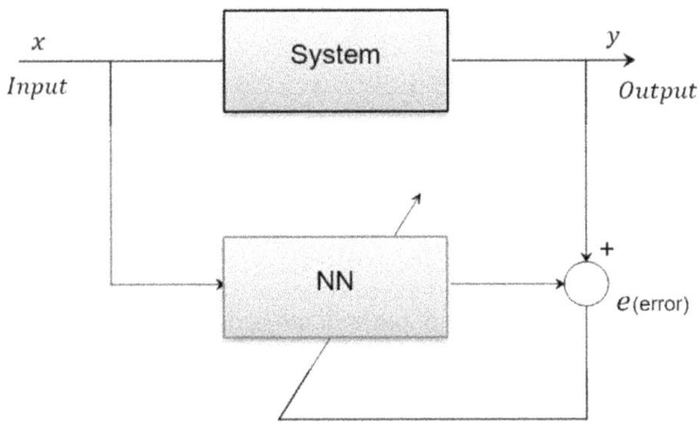

FIGURE 6.5 System modeling by the NN.

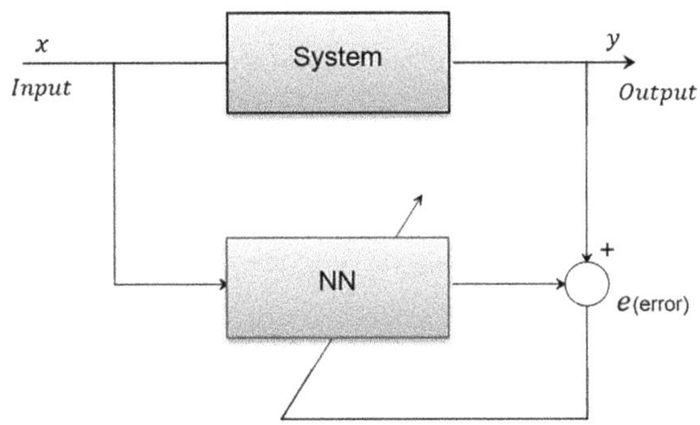

FIGURE 6.6 System Modelling by the NN.

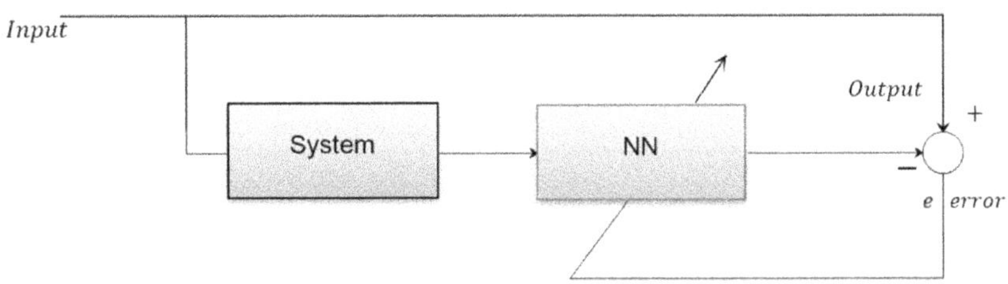

FIGURE 6.7 Inverse system modeling by the NN.

By adding the time delay and noise terms, inverse modeling of the noisy system is shown below (Figure 6.8).

Here, the error term is defined as

$$e[n] = \left| x^*[n] - x[n-1] \right| \tag{6.13}$$

Considering the discrete time $[n]$ (Figures 6.9 and 6.10).

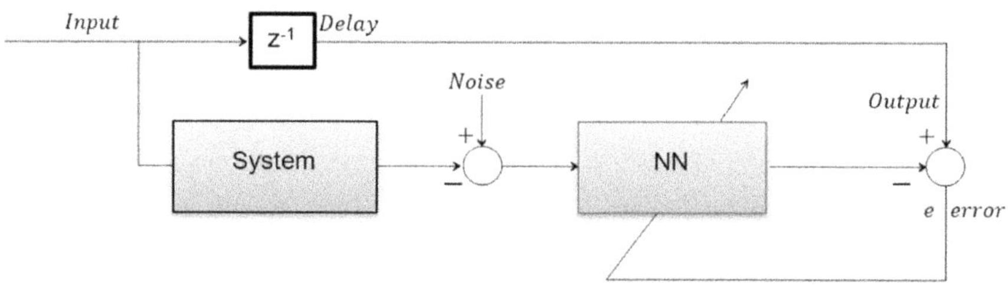

FIGURE 6.8 Noisy inverse modeling.

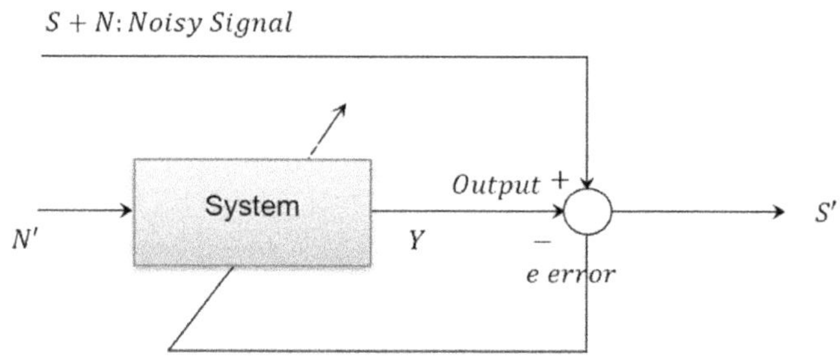

FIGURE 6.9 Noise canceling system.

6.3.3 Adaptive Noise Canceling System

See Figure 6.10.

6.3.4 Adaptive Prediction

Where NN* is a copy of one of the NN used in the inverse modeling, NN* is a slave NN.

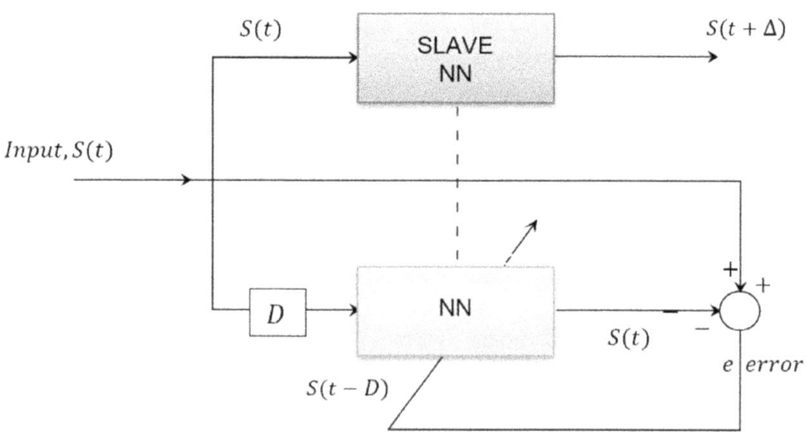

FIGURE 6.10 NN-based adaptive prediction system (D: Delay).

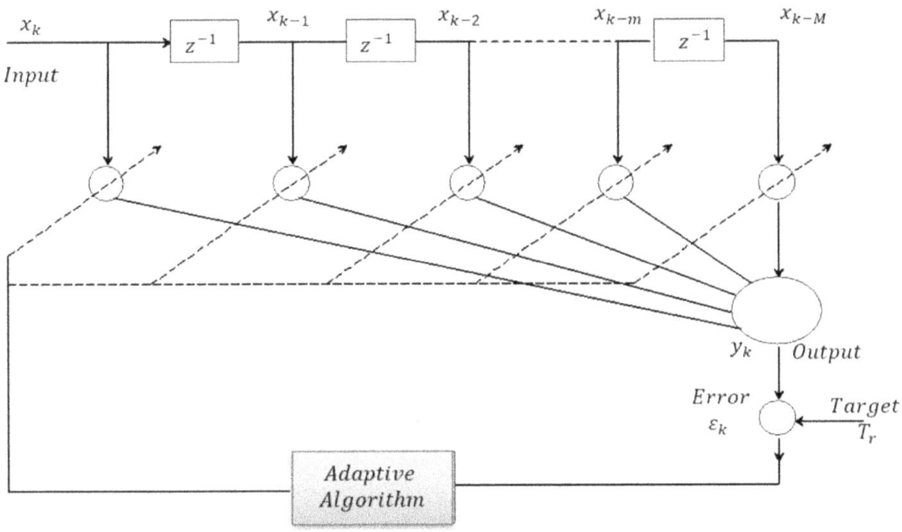

FIGURE 6.11 FIR system.

6.3.5 Non-Parametric Identification

Non-parametric identification uses a Black-Box model defined in linear systems. However, the neural nets within the framework of non-parametric models offer a nonlinear approach to system identification problems. In this manner, Finite Impulse Response (FIR) as a non-parametric system can be implemented by NNs or an adaptive transversal filter as given below: As an example, the adaptive transversal filter structure can be implemented by a NN, and it becomes a FIR network as a non-parametric identification system.

The FIR structure for a given input–output pair, like x and y, is presented by

$$y[n]=\sum_{m=0}^{M}h[m]x[n-m] \tag{6.14}$$

For a single-input adaptive transversal filter with a target output (Figure 6.11).

Here, $y[n]$ is the system output. In the FIR structure $h[m]=b_m$, parameters are represented by the weights in this model, and the adaptive algorithm can adjust these parameters.

6.3.6 Parametric Identification

NNs are trained by means of supervised learning. Also, NNs can be used for both identification and parameter estimation. In this manner, neural nets can be used for parametric identification.

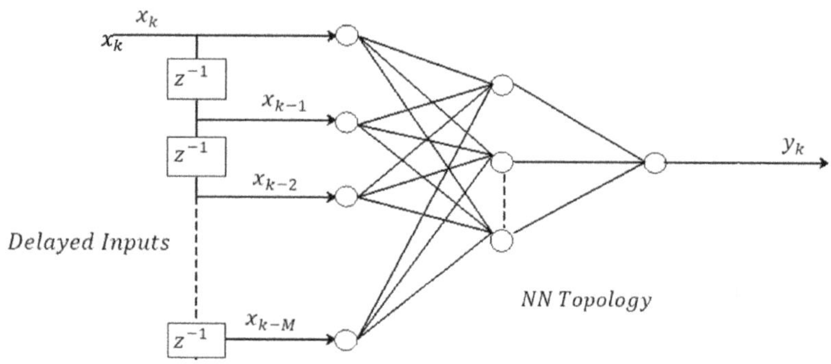

FIGURE 6.12 NN-based FIR system.

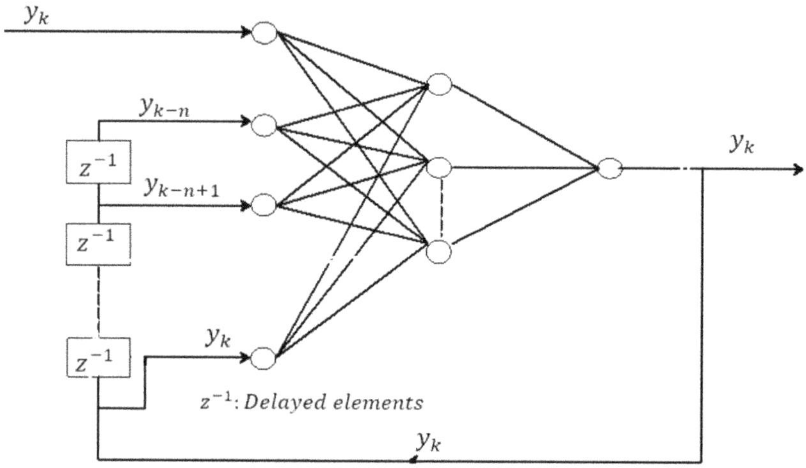

FIGURE 6.13 NN-based IIR System.

Using the advantage of the parametric identification modeling by NNs, the following figures for the Dynamic Model are given for the FIR-filter structure (Figure 6.12).

Sometimes, a dynamical NN with time-delayed recurrent outputs can be used as follows (Figure 6.13):

In this application, the output of the NN is the current value $y[n]$, but it uses the previous values $y[n]$ as inputs. It is used to determine the dependency of current values on former values. This approach is similar to the Infinite Impulse Response (IIR) structure. As a different configuration of the IIR-structure, both of the delayed inputs and outputs can be applied as input information to the neural topology; of course, here, the previous outputs $y[n]$ of the NN are used as feedback connections. For this purpose, the following configuration is given (Figure 6.14).

This is the dynamic NN with time-delayed direct and recurrent inputs, and it provides the Jordan-type RNN topology. From this viewpoint, it is said that Jordan's RNN topology can represent the IIR-filter structure.

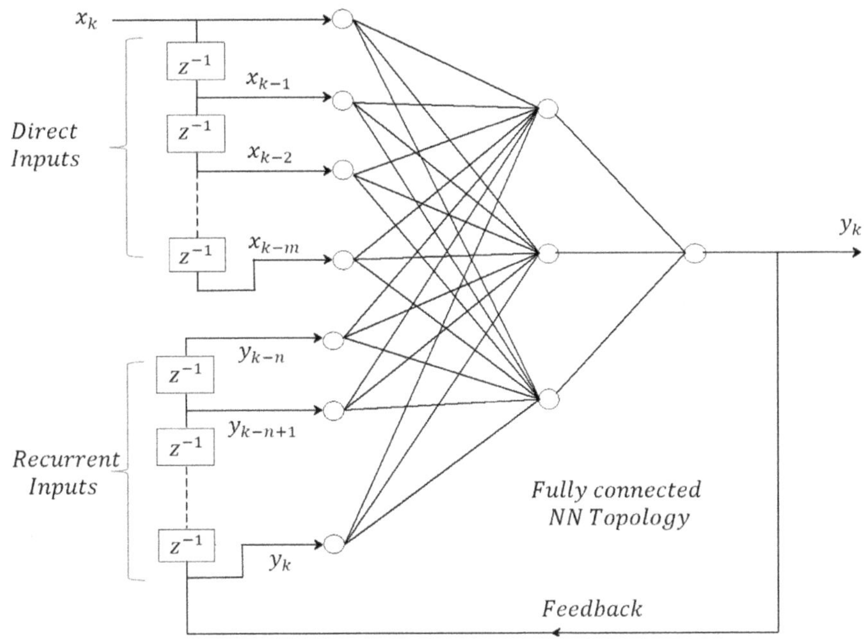

FIGURE 6.14 Dynamic NN with direct and recurrent inputs (IIR System).

6.4 MODELS OF DYNAMICAL SYSTEMS

To get the models of dynamic systems, NN identification structures are used in the configuration of an observer as a connected model to the real physical structure (Figure 6.15).

6.5 CASE STUDIES OF NNs FOR SYSTEM IDENTIFICATION

ANNs have been widely employed in system identification tasks, particularly in fields where nonlinear dependencies and time-series forecasting play a critical role. One such application is wind speed modeling, where ANN-based predictive models can enhance the accuracy of renewable energy forecasting.

In this study, a feedforward ANN model was developed to estimate wind power generation based on hourly wind speed data collected over 1 year in Hatay, a city in southern Türkiye. The ANN was trained using historical wind speed measurements, leveraging the nonlinear relationship between wind speed and power output, which is proportional to the cube of wind speed. The dataset was divided into training and test sets, with 80% of the data used for training and 20% for validation. A single hidden-layer ANN architecture with ten neurons was implemented and trained using the Levenberg–Marquardt optimization algorithm. The model's performance was evaluated using Root Mean Squared Error (RMSE) to assess the accuracy of predictions. The results demonstrated that the ANN model effectively

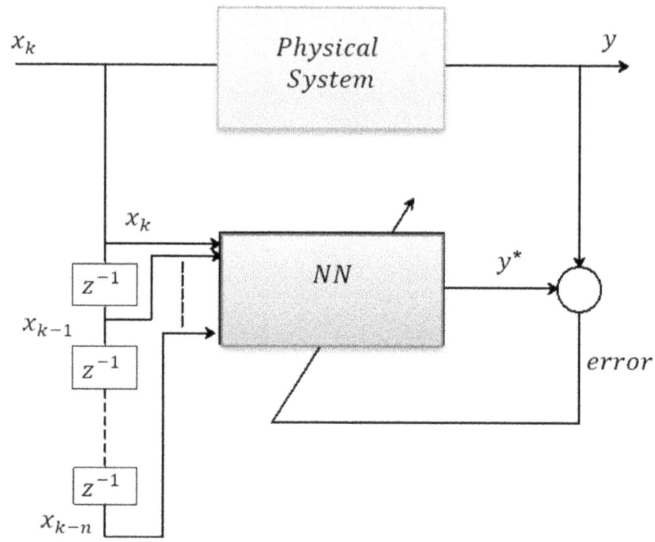

FIGURE 6.15 Non-recurrent parallel identification model.

captured wind speed variations and provided a reliable estimate of wind power output, offering a data-driven approach to improving energy resource planning.

This application highlights the effectiveness of ANN-based system identification in complex, time-dependent environmental systems. By integrating ANN models into renewable energy forecasting, more precise and adaptive energy management strategies can be developed, contributing to improved efficiency in power grid operations.

Figure 6.16 illustrates the hourly wind speed variations recorded in Hatay over the course of 1 year. The graph captures the dynamic and fluctuating nature of the region's wind profile, with speeds ranging between approximately 4 and 16 m/s. These variations highlight the inherent challenges in accurately modeling wind behavior, particularly for renewable energy applications. The dataset presented in this figure was utilized as input for the ANN-based system identification model, showcasing the capability of NNs to analyze and predict complex, time-dependent environmental patterns effectively.

Figure 6.17 compares the measured wind turbine power output with the ANN predictions for the test dataset. The x-axis represents the test samples and the y-axis shows the power output in kilowatts (kW). The measured values are indicated with circle markers, while the ANN predictions are indicated with star markers. The close agreement between the two sets of points indicates that the model effectively captures the nonlinear relationship between wind speed and power generation, providing accurate forecasts in time-varying conditions.

Figure 6.18 presents the learning curve of the artificial neural network (ANN) model over 1,000 epochs. The horizontal axis represents the epoch number, and the vertical axis displays the mean squared error (MSE) on a logarithmic scale, where MSE is the average of the squared differences between the model outputs and the target values. The plot includes the training, validation, and test error curves.

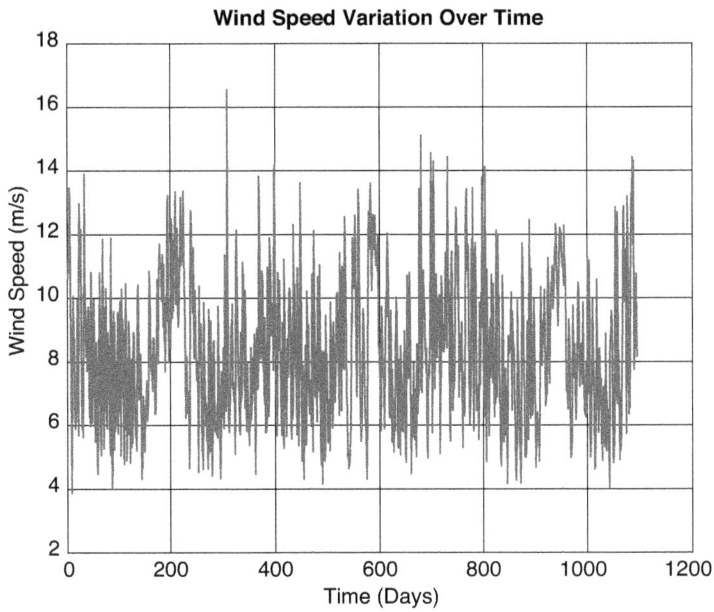

FIGURE 6.16 Hourly wind speed variations in Hatay over 1 year.

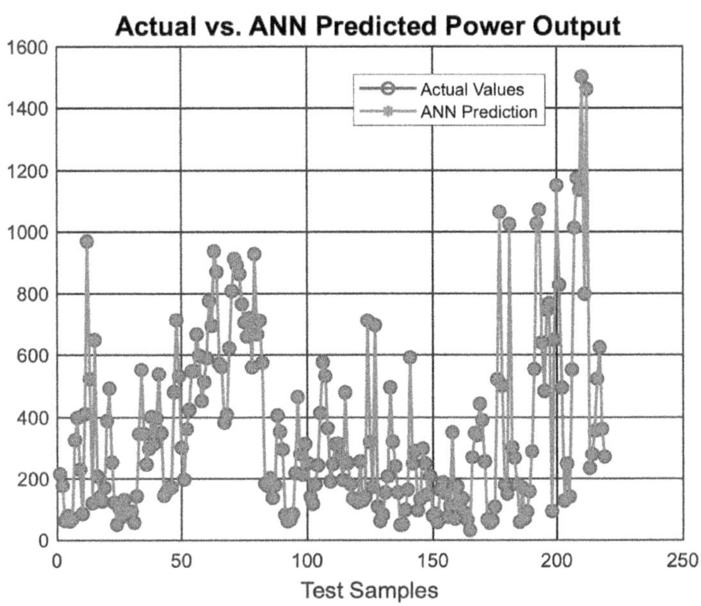

FIGURE 6.17 Comparison of actual power output and ANN predictions.

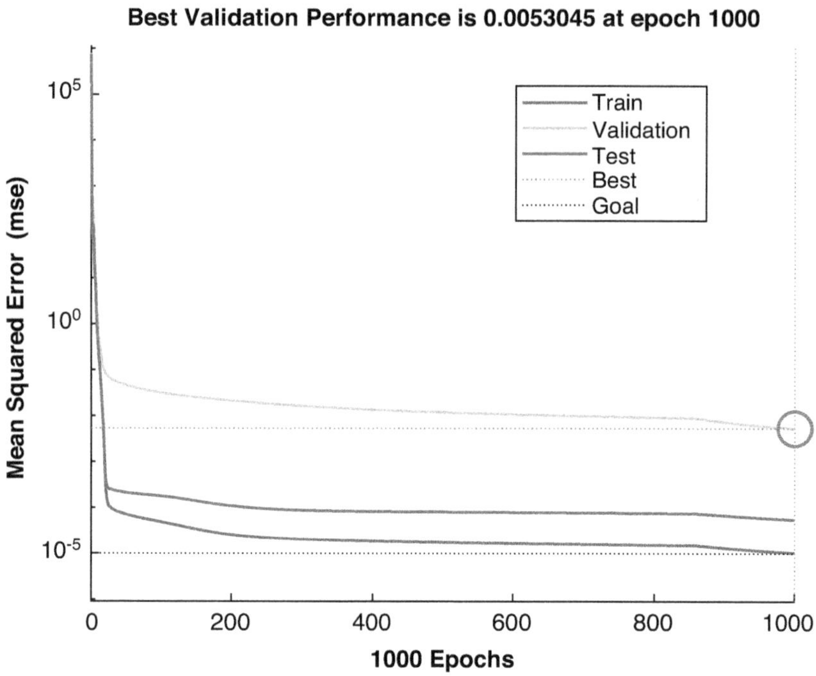

FIGURE 6.18 Training, validation, and test mean squared error across epochs.

The validation curve reaches its minimum at the optimal epoch, indicated by a circle marker, and the test curve follows a similar trend, suggesting that the trained model maintains comparable performance on unseen data. In this figure, the Best and Goal references coincide because the lowest validation error occurs at epoch 1,000 and matches the predefined training objective.

Figure 6.19 presents an error histogram for the ANN model using the training, validation, and test datasets. The x-axis represents the magnitude of the prediction error, and the y-axis indicates the frequency of each error level. Errors are plotted separately for the three data splits, and a zero-error reference line is included for comparison. Most errors are concentrated near zero, indicating that the predictions stay close to the target values. The similar error patterns across training, validation, and test results suggest stable performance without a clear sign of severe overfitting or underfitting.

Figure 6.20 shows regression plots for the ANN model on the training, validation, test, and combined datasets. In each panel, the targets are plotted on the x-axis, and the corresponding model outputs are plotted on the y-axis. The reference line (Y = T) indicates perfect agreement, while the fitted regression line and scattered points summarize the observed performance. Across all panels, the points cluster close to the reference line, and the fitted lines have slopes near 1 with small offsets, indicating low bias and consistent predictive behavior across the different data splits.

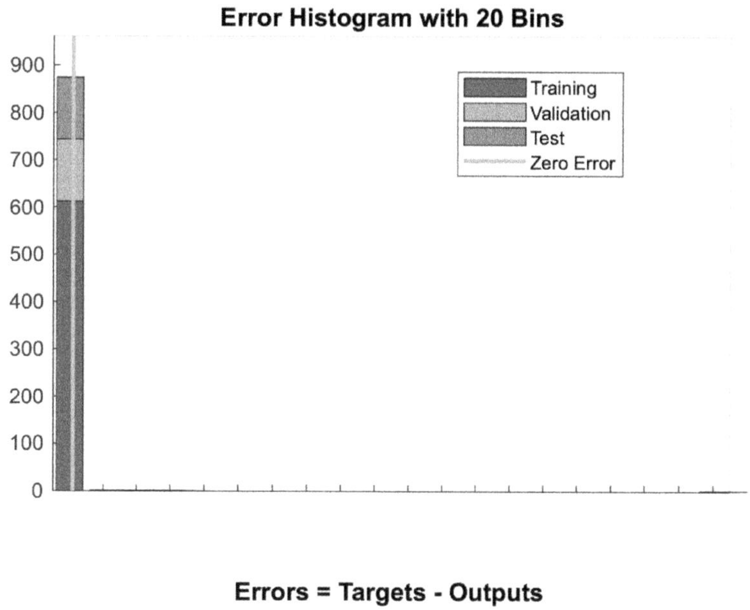

FIGURE 6.19 Error histogram for training, validation, and test data.

The following Python code block is provided to document the workflow used in this study, including data preparation, ANN training, and evaluation for the Hatay wind speed dataset and the associated power output prediction task.

```
# Import required libraries
import numpy as np
import matplotlib.pyplot as plt
from sklearn.model_selection import train_test_split
from sklearn.neural_network import MLPRegressor
from sklearn.metrics import mean_squared_error

# 1. Load the wind speed data from hatay.npy
# Ensure the file contains wind speed data in a NumPy array
wind_speed = np.load('hatay.npy')  # Load data from a .npy
file

# 2. Define variables
# Power output is proportional to the cube of wind speed: P ~
V^3
power_output = 0.5 * wind_speed ** 3

# 3. Visualize Wind Speed Data
time = np.arange(len(wind_speed))  # Create a time vector
corresponding to data length
plt.figure()
```

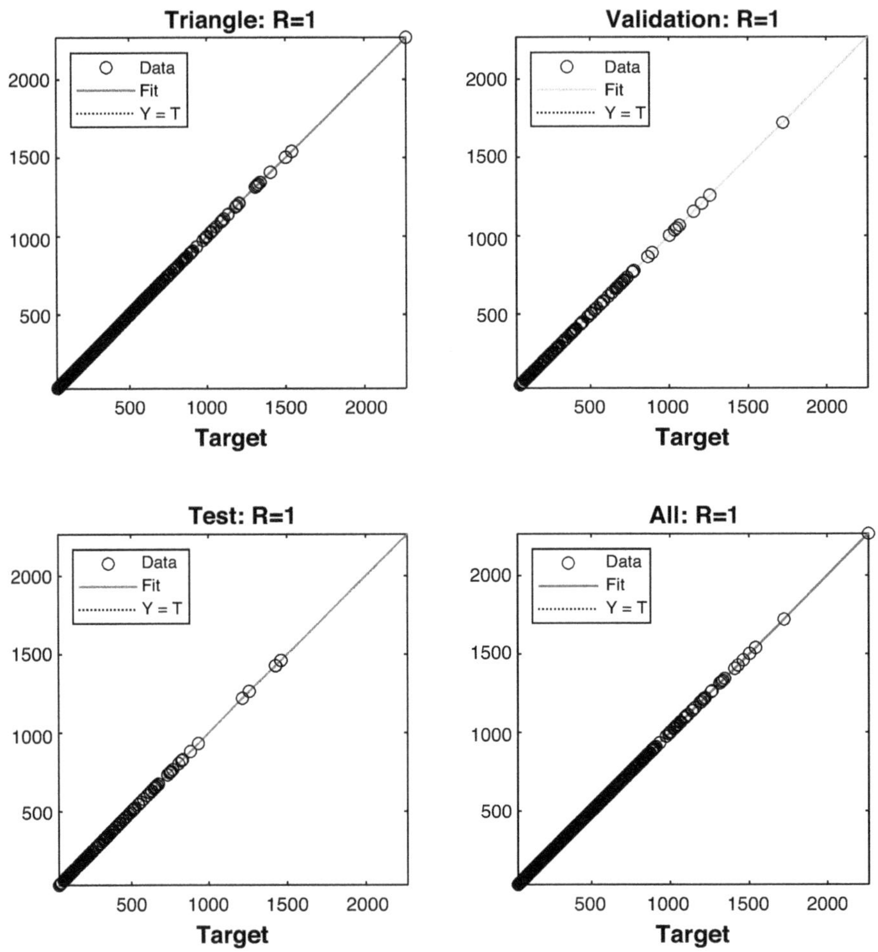

FIGURE 6.20 Regression Plots for Training, Validation, Test, and Overall Data.

```
plt.plot(time, wind_speed, '-b', linewidth=1.5)
plt.grid(True)
plt.xlabel('Time (Days)', fontsize=12)
plt.ylabel('Wind Speed (m/s)', fontsize=12)
plt.title('Wind Speed Variation Over Time (Hatay Dataset)',
fontsize=14)
plt.show()

# 4. Split Data into Training and Testing Sets (80% Training,
20% Testing)
X = wind_speed.reshape(-1, 1)  # Input: Wind speed
Y = power_output  # Output: Power generation
X_train, X_test, Y_train, Y_test = train_test_split(X, Y,
test_size=0.2, random_state=42)
```

```
# 5. Define and Train Artificial Neural Network (ANN) Model
ann = MLPRegressor(hidden_layer_sizes=(10,), max_iter=1000,
learning_rate_init=0.01, tol=1e-5, random_state=42)
ann.fit(X_train, Y_train)

# 6. Test the ANN Model
Y_pred = ann.predict(X_test)

# 7. Visualize Prediction Results
plt.figure()
plt.plot(Y_test, '-ob', linewidth=1.5, label='Actual Values')
plt.plot(Y_pred, '-*r', linewidth=1.5, label='ANN Prediction')
plt.grid(True)
plt.xlabel('Test Samples', fontsize=12)
plt.ylabel('Power Output (kW)', fontsize=12)
plt.title('Actual vs. ANN Predicted Power Output (Hatay
Dataset)', fontsize=14)
plt.legend()
plt.show()

# 8. Compute Prediction Error (RMSE Calculation)
rmse = np.sqrt(mean_squared_error(Y_test, Y_pred))
print(f'Prediction Error (RMSE): {rmse}')
```

NOTE

1 See Appendix B.

7 Neural Network for Applications of Control Theory

7.1 CLASSICAL CONTROLLERS

A classical controller is used to generate the control signal for the controlled system. This means the physical system works in desired responses and stability conditions. A classical block diagram of a control application is given below:

Where,

$r(t)$: reference signal (desired output).
$e(t) = r(t) - y(t)$: error signal.
$u(t)$: control signal to drive the system.

There are three types of controllers:

1. PID Controller.
2. PI Controller.
3. PD Controller.

In a general manner, the PID controller is defined as

PID Controller:

$$u(t) = K_P e(t) + K_I \int e(t)\,dt + K_D \frac{de(t)}{dt} \tag{7.1}$$

where

t: Continuous-time,
K_P, K_I and K_D: Controller parameters for proportional, integral, and derivative, respectively.

Also, other controllers are shown as

PI Controller:

$$u(t) = K_P e(t) + K_I \int e(t)\,dt \tag{7.2}$$

PD Controller:

DOI: 10.1201/9781003673972-7

$$u(t) = K_{P}e(t) + K_{D}\frac{de(t)}{dt} \tag{7.3}$$

Applying the Laplace transform to these equations

For PID:

$$U(s) = K_{P}E(s) + K_{I}\frac{1}{s}E(s) + K_{I}sE(s) \tag{7.4}$$

In a similar way,

For PI:

$$U(s) = K_{P}E(s) + K_{I}\frac{1}{s}E(s) \tag{7.5}$$

And for PD:

$$U(s) = K_{P}E(s) + K_{I}sE(s) \tag{7.6}$$

Here, the most important problem is determining the controller's parameters. One of the most famous methods is 'Ziegler Nichole's' method. Also, use the equalities for each controller type in the '*s*' domain. The transfer functions of the controller are defined:

PID Controller:

$$G_{c}(s) = \frac{U(s)}{E(s)} = K_{P} + K_{I}\frac{1}{s} + K_{I}s \tag{7.7}$$

PI Controller:

$$G_{c}(s) = \frac{U(s)}{E(s)} = K_{P} + K_{I}\frac{1}{s} \tag{7.8}$$

PD Controller:

$$G_{c}(s) = \frac{U(s)}{E(s)} = K_{P} + K_{I}s \tag{7.9}$$

Also, considering the transfer function of the physical system, like $G_{s}(s)$, the first block diagram that is given in Figure 7.1 becomes Figure 7.2.

Here $R(s)$ and $Y(s)$ are also representation $r(t)$ (reference signal) and $y(t)$ (system output) in the *s*-domain, respectively.

Next, the controller parameters are determined using the Ziegler–Nichols tuning method for the PID controller, as follows:

$$K_{P} = 0.6K_{m} \tag{7.10}$$

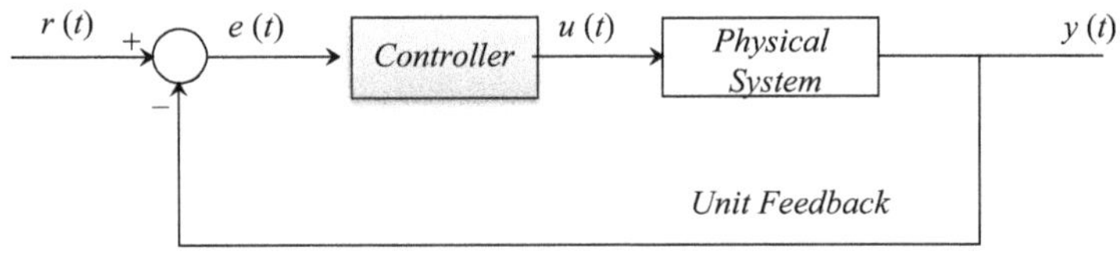

FIGURE 7.1 A simple control system with unit feedback.

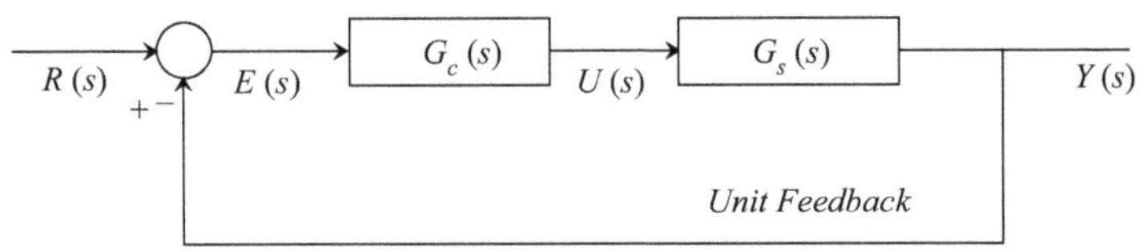

FIGURE 7.2 Simple control system in the s-domain with unit feedback.

$$K_I = \frac{K_P \omega_m}{\pi} \tag{7.11}$$

$$K_D = \frac{K_P \pi}{4\omega_m} \tag{7.12}$$

where K_m is the proportional gain of the oscillating system and ω_m is the oscillation frequency.

Either the root locus or the Bode plots can be used to determine K_m and ω_m.

For example, a root locus is obtained from the plant transfer function. The gain at which the root locus crosses the *j*-axis is K_m, and the frequency on the $j-\omega$ axis gives ω_m.

Alternatively, Bode plots are plotted for the given plant (physical system) transfer function. The G_m (Gain Margin) is determined at the frequency ω_{pc}:

$$K_m = 10^{\left(G_m/20\right)} \tag{7.13}$$

$$\omega_m = \omega_c \tag{7.14}$$

Example 7.1

For a given plant,

$$G(s) = \frac{400}{s\left(s^2 + 20s + 240\right)}$$

Find the PID Controller parameters by the Ziegler–Nichols Method.

Solution:

Using the following MATLAB commands:

```
s = tf('s');
g=400/(s*(s^2+20*s+240));
sisotool(g)
```

Some system parameters are

Closed-loop poles $= \{-4.2 \pm j0.932, -21.6\}$

Gain Cross-over frequency = 1.95 rad/s

GM (Gain Margin) = 23 dB

PM (Phase Margin) = 1.95°.

The parameters using Equations (7.13) and (7.14) are $K_m = 14.1254$ and $\omega_m = 14.1 \frac{\text{rad}}{\text{s}}$, respectively.

PID parameters using Equations (7.10)–(7.12) are $K_P = 8.47$, $K_I = 38.015$, and $K_D = 2.12$, respectively.

The closed-loop step response has no overshoot, and the steady-state error to a unit ramp input is 0.5. The root locus and Bode plots for this plant are given in Figures 7.3 and 7.4.

These three controller parameters affect the controller's performance. The effects of each controller K_P, K_I, and K_D on a closed system are summarized as shown in Table 7.1.

Here, the system response indicators like rise time, overshoot, settling time, and steady-state error, depending on the performance of the controller, are defined

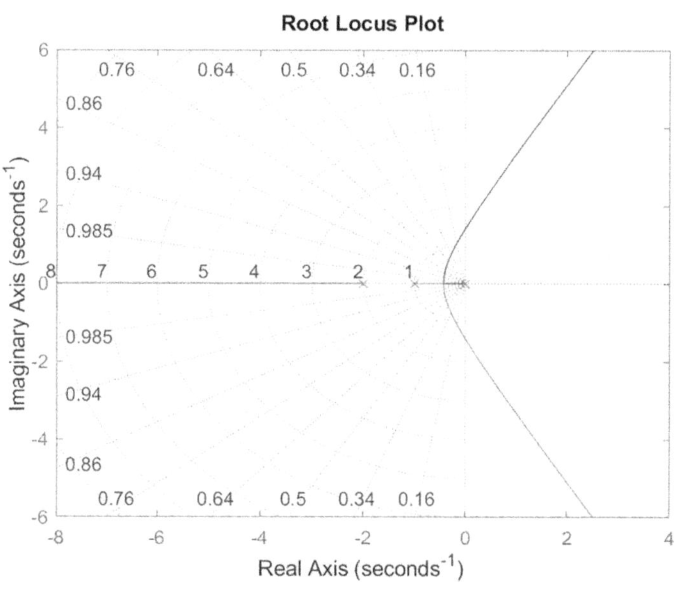

FIGURE 7.3 Root locus plot.

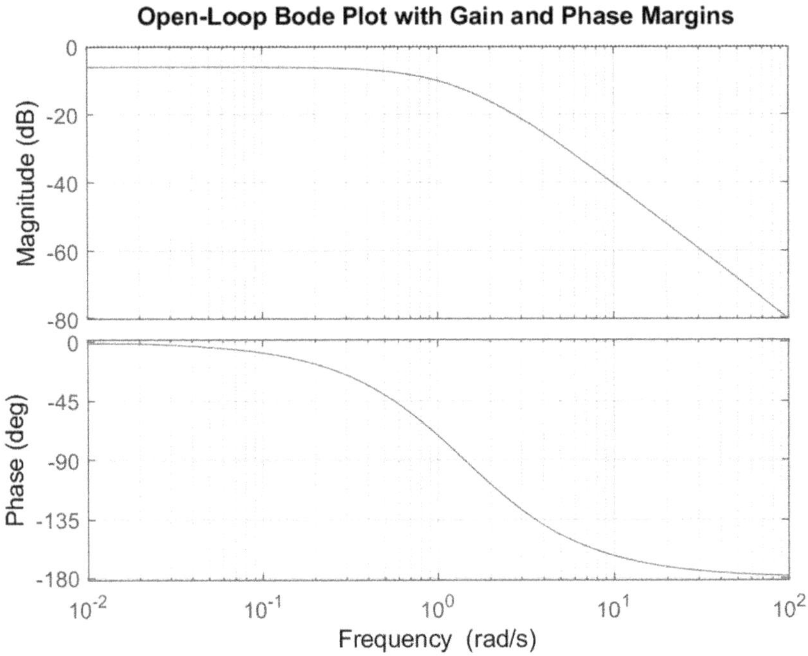

FIGURE 7.4 Bode plots.

TABLE 7.1
Second-Order System Response to a Unit Step and Its Control Parameters

	Rise Time	Overshoot	Settling Time	Steady-State Error
K_P	Decrease	Increase	Small change	Decrease
K_I	Decrease	Increase	Increase	Eliminate
K_D	Small change	Decrease	Decrease	Small change

by the following figure of the second-order system response to the unit input (Figure 7.5).

$t_{r:}$ rise time.
$t_{p:}$ peak time.
$t_{s:}$ setting time.

NEURO CONTROLLERS

The computation of the controller's parameters can be challenging, and to address this difficulty, the neural network approach offers an alternative solution for control applications. In this sense, neuro-controller structures are defined.

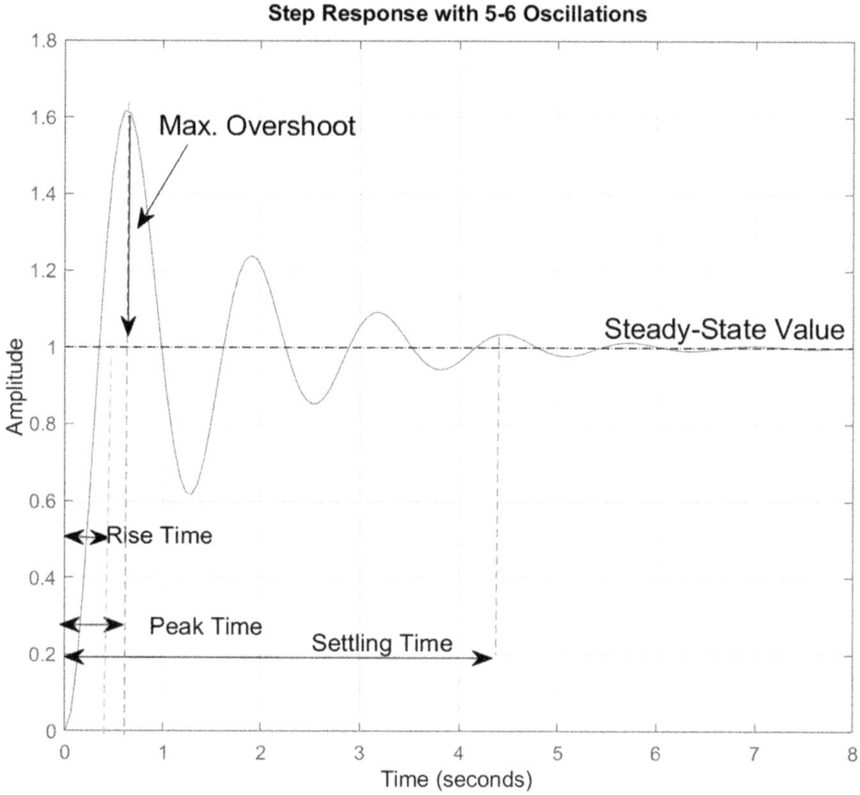

FIGURE 7.5 Second-order system response to a unit step function.

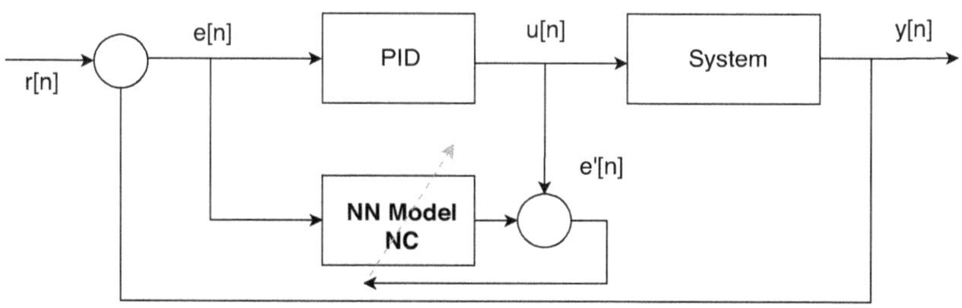

FIGURE 7.6 NN-based observer for the PID controller.

Here, the neuro-controller structures approach is a learning system that is independent of the controller parameter calculation. As a primitive (basis) model (Figure 7.6).

Here, we have the PID controller structure, and training the Neural Networks (NN) with the input and output (target for the NN) pairs of the PID controller in the NN topology serves as the role of the classical PID controller. This application is not independent of the classical controller structure, but it provides the controller with reliability. It can be accepted as a redundant system.

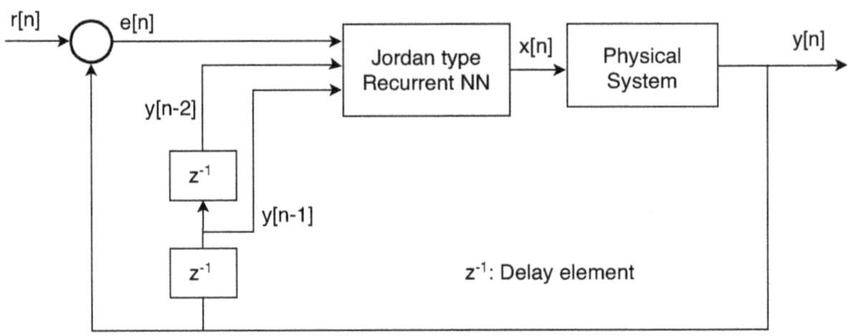

FIGURE 7.7 Direct use of the NN as a controller.

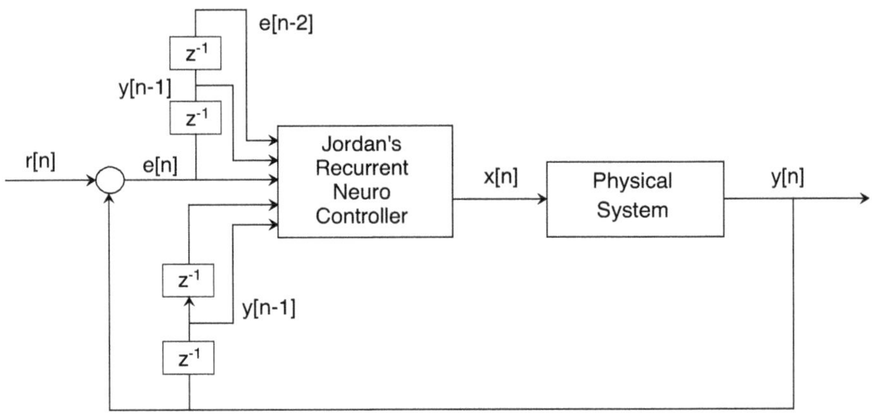

FIGURE 7.8 Direct use of an RNN-based controller with a delayed error sequence.

7.2 NEURO-CONTROLLER APPLICATIONS

As shown in the following figure, it is of direct use (Figure 7.7).

However, it requires batched training in conjunction with the physical system. For this reason, its training is somewhat more challenging compared to the training of NNs alone (Figure 7.8).

In application, a mathematical model of the physical system exists.

The training of the NN unit to assume the controller role as a compact system with a mathematical model of the system can be utilized. It can be prepared as a neuro-controller. Additionally, its software can be utilized in the form of an embedded system, allowing a neuro-controller to be designed as a physical unit in real-world applications.

7.3 SELF-TUNING PID CONTROLLER BY NEURAL NETWORKS

Rather than using the Neural Controller alone, a hybrid approach can be employed in which the PID controller is integrated with the neural network structure. This

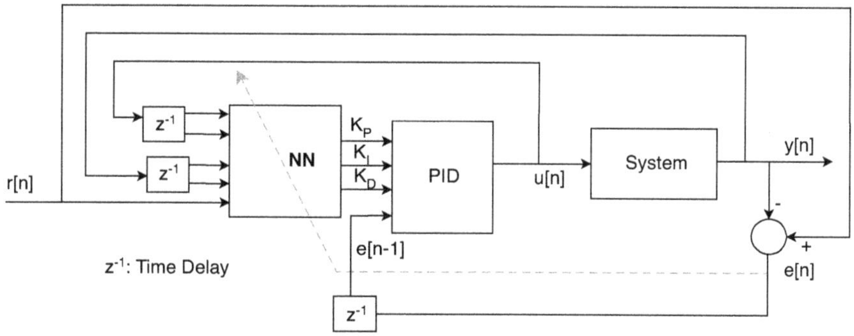

FIGURE 7.9 Self-tuning PID controller by the NN.

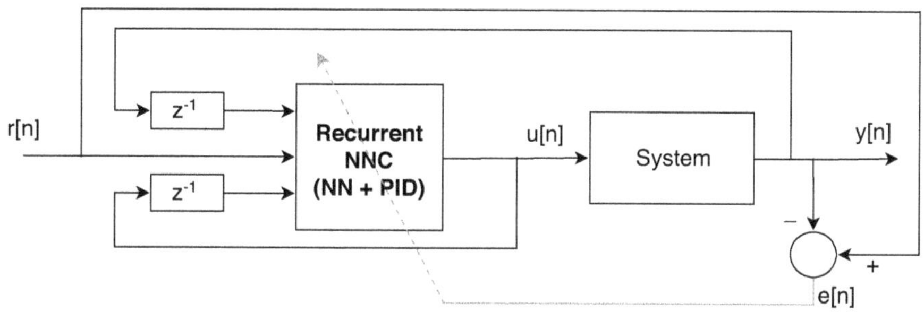

FIGURE 7.10 Direct use of RNN in system control.

integration results in a self-tuning PID controller driven by the neural network, as illustrated below. Here, *TD* represents the time delay (Figure 7.9).

However, combining the PID and NN units is unique. The direct use of the NN can be rebuilt in a new form. This provides an alternative use independent of the PID parameters (Figure 7.10).

7.4 COMPARISONS: NEURAL NETWORK-BASED CONTROL vs. PID CONTROL FOR DYNAMIC SYSTEMS

Control systems play a crucial role in engineering applications, where maintaining stability and desired performance is essential. In this section, we compare PID control and ANN-based control using a synthetically generated dynamic system. The goal is to observe the response of both controllers in maintaining system stability and minimizing error.

A second-order dynamic system is selected for this study, and a synthetic dataset is generated to represent its behavior under different conditions. We implement both a classical PID controller and a neural network-based adaptive controller, evaluating their effectiveness in trajectory tracking and disturbance rejection (Figure 7.11).

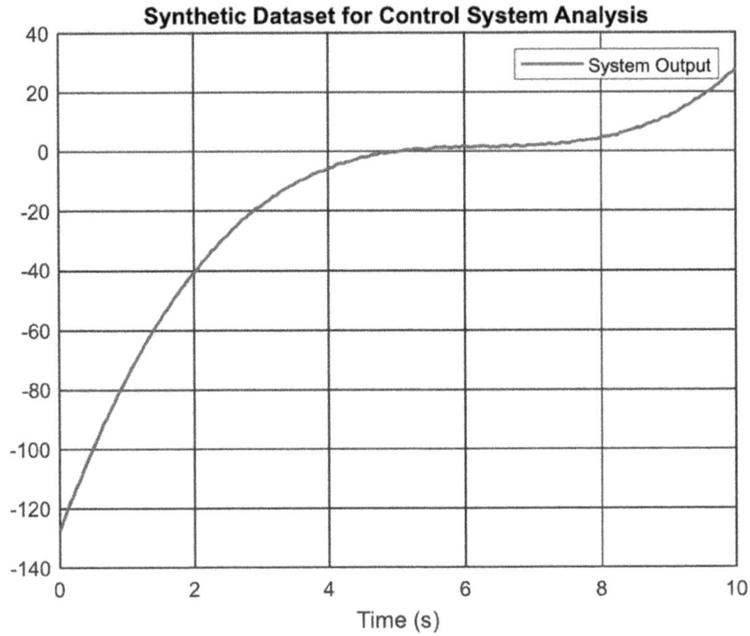

FIGURE 7.11 Open-loop system response.

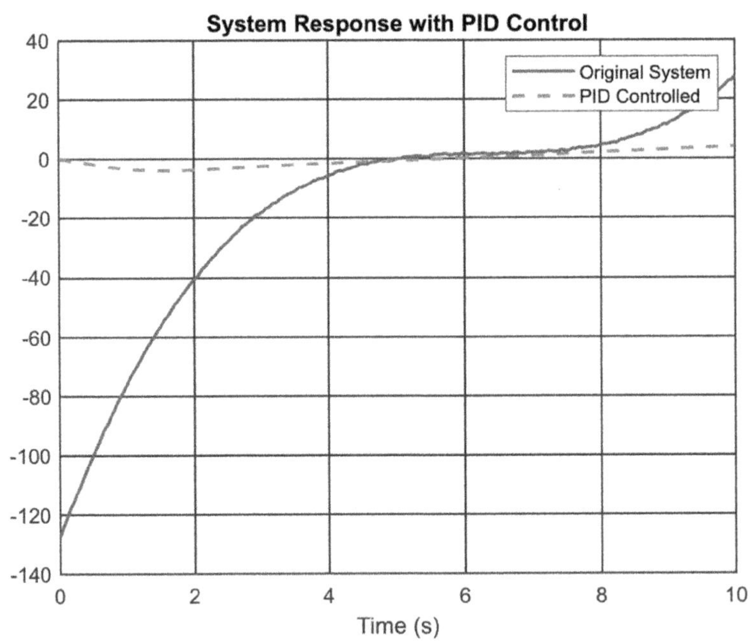

FIGURE 7.12 System response with PID control.

This figure illustrates the system's response without any control mechanism applied. The curve represents the system output, which follows its natural trajectory. The lack of control results in slow stabilization and potential deviations from the desired output, highlighting the need for an effective control strategy (Figure 7.12).

In this figure, a PID controller is implemented to regulate the system's behavior. The curve shows the original system response, while the red dashed line represents the PID-controlled output. The PID controller effectively reduces steady-state error and helps stabilize the system by mitigating oscillations and improving convergence speed.

```
import numpy as np
import matplotlib.pyplot as plt
from scipy.integrate import odeint

# Step 1: Generate Synthetic Dataset
np.random.seed(42)
t = np.linspace(0, 10, 200)  # Time vector
x = np.linspace(-5, 5, 200)
noise = np.random.randn(len(x)) * 0.2
y = 0.5 * x**3 - 2 * x**2 + 3 * x + noise  # Synthetic dataset
u = y  # System input

# Step 2: Define Second-Order System Model
def system_model(y, t, u):
    dydt = [y[1], -2*y[1] - y[0] + u[int(t / (t[-1] / len(t)))
% len(t)]]
    return dydt

# Simulate Open-Loop Response
y0 = [0, 0]
y_response = odeint(system_model, y0, t, args=(u,))[:, 0]

# Step 3: Implement PID Controller
Kp, Ki, Kd = 3, 1, 0.5
y_pid = np.zeros_like(y)
prev_error, integral = 0, 0
dt = t[1] - t[0]

def pid_controller(error, prev_error, integral, dt):
    derivative = (error - prev_error) / dt
    integral += error * dt
    return Kp * error + Ki * integral + Kd * derivative,
integral

for i in range(1, len(y)):
    error = u[i] - y_pid[i-1]
    control_signal, integral = pid_controller(error, prev_
error, integral, dt)
    y_pid[i] = y_pid[i-1] + control_signal * dt
    prev_error = error
```

```
# Plot 1: Open-Loop System Response
plt.figure()
plt.plot(t, y_response, 'b', linewidth=1.5)
plt.grid()
plt.xlabel('Time (s)')
plt.ylabel('System Response')
plt.title('Figure 1: Open-Loop System Response')
plt.legend(['System Output'])
plt.show()

# Plot 2: System Response with PID Control
plt.figure()
plt.plot(t, y_response, 'b', linewidth=1.5)
plt.plot(t, y_pid, 'r--', linewidth=1.5)
plt.grid()
plt.xlabel('Time (s)')
plt.ylabel('System Output')
plt.title('Figure 2: System Response with PID Control')
plt.legend(['Original System', 'PID Controlled'])
plt.show()
```

8 Condition Monitoring with Neural Nets

8.1 SENSOR VALIDATION

In system-level applications, ensuring the reliability of sensor data is essential for safe and accurate operation. Faulty or inconsistent sensor readings can lead to incorrect conclusions, making sensor validation a critical part of condition monitoring. Consider, for instance, a scenario involving three sensors labeled A, B, and C, each responsible for monitoring different system parameters.

Figure 8.1 shows how an auto-associative NN is trained using input–output pairs from all system sensors (e.g., *x–a*, *y–b*, and *z–c*). In the event of an anomaly, the NN produces a noticeable error at its output. This discrepancy enables the detection of faults by comparing the NN's output with the actual sensor readings. The approach relies on analyzing these deviations in the time domain for effective error identification.

8.2 CONDITION MONITORING APPLICATION OF THE NN AS A NERO-DETECTOR

While most neural network applications focus on time-domain signals, identifying faults often requires analysis in the frequency domain. Therefore, as illustrated in Figure 8.2, the spectra of time-domain signals are computed and used as inputs to the neural network. The spectral data represent the amplitude spectrum of the original signal, allowing the network to process frequency-domain characteristics. This approach enhances the network's ability to detect anomalies by capturing more detailed signal behavior beyond what is visible in the time domain.

Different system behaviors result in varying levels of error signals. These variations reveal distinct frequency components in the spectral domain. Figure 8.3

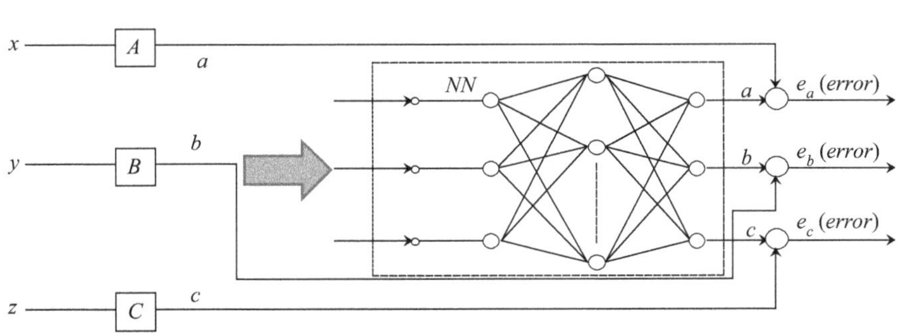

FIGURE 8.1 Sensor monitoring by an auto-associative NN.

DOI: 10.1201/9781003673972-8

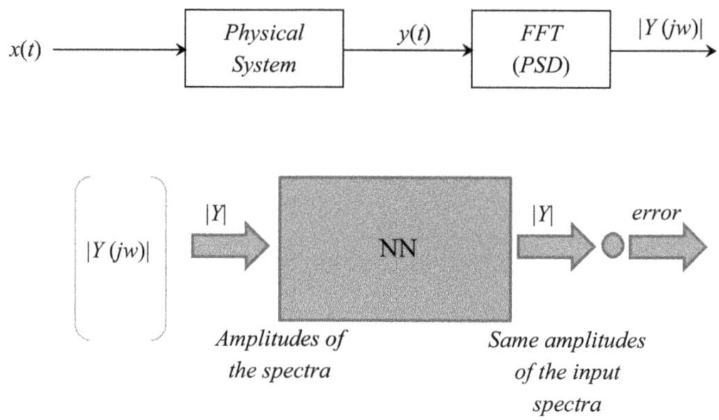

FIGURE 8.2 Auto-associative neuro detector.

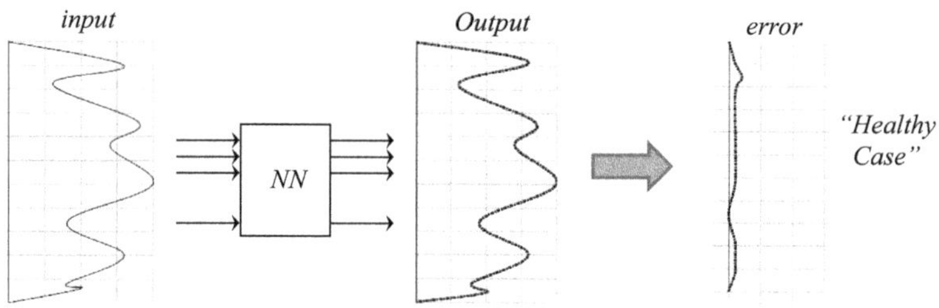

FIGURE 8.3 Normal condition with a small error change.

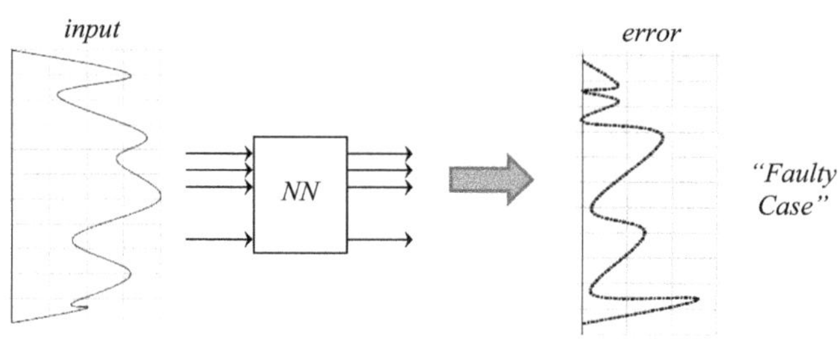

FIGURE 8.4 Fault detection by error change.

illustrates the spectral characteristics under a normal operating condition, where the error changes remain minimal, and the system appears stable.

When the neural networks use the same configuration and training procedure, the system response under the first healthy operating condition can be taken as a reference (see Figure 8.4). In this case, the neural network (NN) output error varies only slightly. Under faulty conditions, the error spectrum shows pronounced amplitude changes, typically appearing as distinct peaks. The dominant peak frequency can be extracted from the spectrum and used to identify the corresponding fault, since different faults introduce characteristic frequency components.

Appendix A
Random Noise and Statistical Parameters

Random noise is not merely a theoretical construct but a critical element in modeling measurement uncertainty and environmental disturbances in real-world systems. It enables researchers to test the robustness of algorithms and simulate unpredictable behavior in signals, which is crucial for designing effective filtering and fault detection methods. By analyzing statistical parameters such as the mean, variance, skewness, and kurtosis, one can gain a deeper understanding of the nature of noise and its impact on signal processing performance. This section introduces the foundational tools needed to quantify and interpret random noise in system applications.

A.1 GAUSSIAN NOISE

Gaussian noise is widely used in signal processing and system simulations to model random fluctuations and uncertainties that naturally occur in real-world environments. It allows engineers and researchers to test the robustness and performance of algorithms under realistic, noisy conditions. This type of noise is essential in validating system responses and designing filters or learning models that can tolerate unpredictable variations in input data.

The expression for Gaussian Noise is given by the help of the normal distribution with μ mean value and σ^2 variance:

$$\text{Gaussian Noise} = k{*}N\ \left(\mu,\sigma^2\right)$$

where

k: Gain Factor that scales the amplitude of the raw random data. It is proportional to the standard deviation of the random signal. A general expression for generating such a noise sequence can be written in vector form as follows:

$$x = r\,\text{and}\,n(1,N).$$

Here, N is the number of generated random numbers and k is a gain factor that scales the amplitude of the raw random data and is proportional to the standard deviation of the signal. In practical terms, this formulation generates a sequence of random values drawn from a standard normal distribution $\left(N(0,1)\right)$ namely where typically $\mu = 0$ and $\sigma^2 = 1,$ resulting in a zero-mean, unit-variance noise profile commonly used in signal processing applications.

A.2 STATISTICAL PARAMETERS

Statistical parameters are crucial for analyzing and interpreting the behavior of signals, particularly in systems subject to uncertainty or dynamic changes. These parameters help quantify central tendencies, variability, and the overall shape of a signal's distribution. In real-world applications, such as control systems, diagnostics, or signal filtering, they are used to detect anomalies, assess signal stability, and support decision-making under uncertain conditions. The following are key metrics used to characterize signal behavior:

Mean (μ): The mean, as defined in Equation (A.1), represents the average value of a signal over a specific observation period. It serves as a baseline around which the signal fluctuates and provides a reference point for interpreting whether values deviate consistently or erratically. In real-world applications, such as environmental monitoring or power system measurements, the mean is used to determine nominal operating conditions. For example, in temperature control systems, maintaining a stable mean temperature is crucial for system efficiency and safety. The mean helps identify shifts in signal behavior and, when tracked over time, can reveal long-term trends or gradual drifts in a system's performance:

$$\mu = \frac{1}{N}\sum_{i=1}^{N} x_i \tag{A.1}$$

Standard Deviation (σ): The standard deviation, shown in Equation (A.2), quantifies how much individual values in a signal deviate from the mean. It is a direct measure of the signal's variability or noise level. A low standard deviation indicates that the signal remains close to the average, reflecting stability—a common characteristic of well-regulated control systems. In contrast, a high standard deviation implies increased fluctuation or instability, which may indicate disturbances or faults. In experimental settings, standard deviation is essential for defining confidence intervals and measurement uncertainty, allowing engineers and researchers to assess how reliable and repeatable their results are:

$$\sigma = \sqrt{\frac{1}{N}\sum_{i=1}^{N} \left(x_i - \mu\right)^2} \tag{A.2}$$

Variance (σ^2)**:** Variance, described in Equation (A.3), is the square of the standard deviation and provides a broader view of signal dispersion. While standard deviation is easier to interpret in terms of scale, variance is often used in statistical modeling and simulation, such as in stochastic processes or Monte Carlo analyses. In practice, variance helps distinguish between normal variability and abnormal conditions. For instance, in the vibration analysis of rotating machinery, a sudden increase in variance may signal mechanical wear or imbalance. Variance is particularly helpful in comparing the consistency of different systems or sensors over time:

$$\sigma^2 = \frac{1}{N}\sum_{i=1}^{N}(x_i - \mu)^2 \tag{A.3}$$

Skewness (α): Skewness, as defined in Equation (A.4), measures the asymmetry of the signal distribution. A perfectly symmetric signal has a skewness of zero. A positive skew indicates that the signal has a longer tail on the right side, suggesting that more frequent occurrences of low values are accompanied by occasional high spikes. Conversely, a negative skew shows a longer tail on the left side. In real-world systems, skewness is useful for identifying persistent bias in sensor outputs or financial signals. For example, in load forecasting, a right-skewed demand profile may indicate occasional peak loads, which are important for system planning and energy pricing strategies:

$$\alpha = \frac{1}{N}\sum_{i=1}^{N}\left(\frac{x_i - \mu}{\sigma}\right)^3 \tag{A.4}$$

Kurtosis (β): Kurtosis, given in Equation (A.5), evaluates the 'tailedness' or sharpness of a signal's distribution. A high kurtosis value indicates the presence of extreme values or outliers, which may result from transient events, anomalies, or faults. Low kurtosis suggests a flatter distribution with fewer outliers. In diagnostic applications, such as fault detection in transformers or engines, high kurtosis can reveal rare but critical abnormalities that the mean or variance might overlook. It is especially useful in condition-based monitoring systems, where early detection of abnormal signal behavior can prevent costly failures:

$$\beta = \frac{1}{N}\sum_{i=1}^{N}\left(\frac{x_i - \mu}{\sigma}\right)^4 \tag{A.5}$$

Appendix B
Linear Stochastic Models

Linear stochastic models offer a concise and practical approach to describing time-dependent signals that exhibit both a deterministic structure and random fluctuations. They are widely used in signal processing and time-series analysis because they enable (1) prediction from past observations, (2) interpretable parameter estimation, and (3) principled model-order selection. In many engineering and scientific applications, these models form a baseline for forecasting, system identification, and decision-making under uncertainty.

B.1 AUTOREGRESSIVE MODEL

In time-series analysis, autoregressive (AR) stands for autoregressive. An Auto-Regressive (AR) process models the current sample as a linear combination of its past samples plus a random innovation term. The most common AR model of signal y with order p is written as:

$$y_t = c + \sum_{i=1}^{p} \phi_i \, y_{t-i} + \varepsilon_t \tag{B.1}$$

where y_t is the observed signal, c is an optional constant term, ϕ_i are the AR coefficients, and ε_t is a zero-mean white-noise (innovation) sequence with variance σ_ε^2. Model (B.1) is widely used for one-step-ahead prediction and for describing signals whose short-term dynamics are dominated by their recent history.

From a signal-processing viewpoint, the AR model corresponds to an all-pole IIR filter driven by white noise. In the z-domain, the transfer function from ε_t to y_t (ignoring c for simplicity) is:

$$H(z) = \frac{1}{1 - \sum_{i=1}^{p} \phi_i z^{-i}} \tag{B.2}$$

B.1.1 Stationarity condition

For the AR(p) process to be wide-sense stationary, the roots of the characteristic polynomial must satisfy a stability condition. A standard form uses:

$$A(z) = 1 - \phi_1 z^{-1} - \phi_2 z^{-2} - \cdots - \phi_p z^{-p} \tag{B.3}$$

The process is stationary if all roots of $A(z) = 0$ lie outside the unit circle. In particular, AR(1) with $y_t = c + \phi_1 y_{t-1} + \varepsilon_t$ is stationary if $|\phi_1| < 1$.

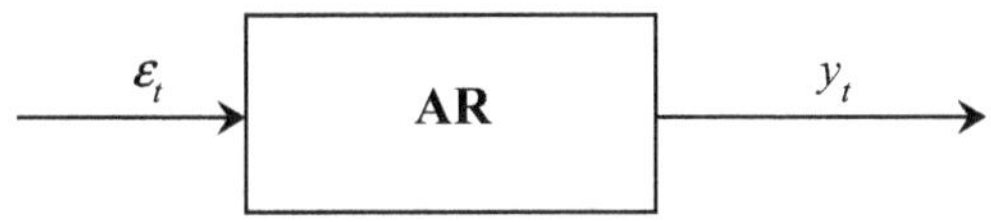

FIGURE B.1 AR model.

B.1.2 Relation to MA and ARMA

An AR model depends on past outputs $\{y_{t-i}\}$. A moving-average (MA) model depends on past innovations $\{\varepsilon_{t-i}\}$. An ARMA model combines both past outputs and past innovations, providing a more flexible description when pure AR or pure MA structures are insufficient.

B.2 YULE-WALKER (AUTOCORRELATION) ALGORITHM

A classical approach for estimating AR coefficients is based on the Yule–Walker equations, which connect AR parameters to the autocorrelation of the process. For clarity, assume a zero-mean process (or equivalently, that the mean has been removed). Let:

$$r(k) = \mathbb{E}\{y_t \, y_{t-k}\} \tag{B.4}$$

denote the autocorrelation at lag k.

Starting from (B.1) (with $c = 0$ for simplicity), multiply both sides by y_{t-d} for $d \geq 1$ and take expectations:

$$\mathbb{E}\{y_t y_{t-d}\} = \sum_{i=1}^{p} \phi_i \, \mathbb{E}\{y_{t-i} y_{t-d}\} + \mathbb{E}\{\varepsilon_t y_{t-d}\} \tag{B.5}$$

Because ε_t is white noise and uncorrelated with past samples $\{y_{t-d}\}$ for $d \geq 1$, we have $\mathbb{E}\{\varepsilon_t y_{t-d}\} = 0$. Using (B.4), (B.5) becomes:

$$r(d) = \sum_{i=1}^{p} \phi_i \, r(d-i), d = 1,2,\ldots,p \tag{B.6}$$

Equation (B.6) can be written in matrix (Toeplitz) form as:

$$\begin{bmatrix} r(0) & r(1) & \cdots & r(p-1) \\ r(1) & r(0) & \cdots & r(p-2) \\ \vdots & \vdots & \ddots & \vdots \\ r(p-1) & r(p-2) & \cdots & r(0) \end{bmatrix} \begin{bmatrix} \phi_1 \\ \phi_2 \\ \\ \phi_p \end{bmatrix} = \begin{bmatrix} r(1) \\ r(0) \\ \\ r(p) \end{bmatrix}. \tag{B.7}$$

In practice, $r(k)$ is replaced by its sample estimate computed from a finite record of M samples, and the AR coefficients are obtained by solving the linear system (B.7). Efficient solvers exist due to the Toeplitz structure (e.g., Levinson–Durbin recursion).

B.3 BURG ALGORITHM

Another widely used parameter-estimation method is Burg's method. Unlike Yule–Walker estimation, which relies on estimated autocorrelations, Burg's method directly minimizes the forward and backward prediction errors while enforcing a stable AR model. It estimates reflection (partial autocorrelation) coefficients through a recursion related to the Levinson–Durbin algorithm, and then converts them to AR coefficients.

For an AR(p) model, the one-step-ahead predictor is:

$$\hat{y}_t = c + \sum_{i=1}^{p} \phi_i \, y_{t-i} \tag{B.8}$$

and the (forward) prediction error is defined as:

$$e_t = y_t - \hat{y}_t \tag{B.9}$$

Burg's method chooses the parameters so that the sum of squared forward and backward errors is minimized at each order, yielding accurate estimates, particularly for short data records, and often producing stable spectra.

The prediction-error variance provides a compact measure of model fit. Given M samples and AR order p, a standard estimate is:

$$\hat{\sigma}_e^2 = \frac{1}{M-p} \sum_{t=p+1}^{M} e_t^2 \tag{B.10}$$

where e_t is defined in (B.9). This estimate is used both for reporting model accuracy and for order selection criteria.

B.4 AKAIKE'S INFORMATION CRITERIA (AIC)

Selecting the model order p is a key practical step: a too-small order may miss important dynamics, while a too-large order may overfit noise. The Akaike Information Criterion (AIC) provides a principled balance between goodness of fit and complexity. For an AR(p) model fitted to M samples with (approximately) Gaussian innovations, a commonly used form is:

$$\text{AIC}(p) = M \ln\left(\hat{\sigma}_e^2(p)\right) + 2p \tag{B.11}$$

where $\hat{\sigma}_e^2(p)$ is the prediction-error variance estimate obtained for order p (e.g., (B.10)). The preferred order is the value p that minimizes $\text{AIC}(p)$.

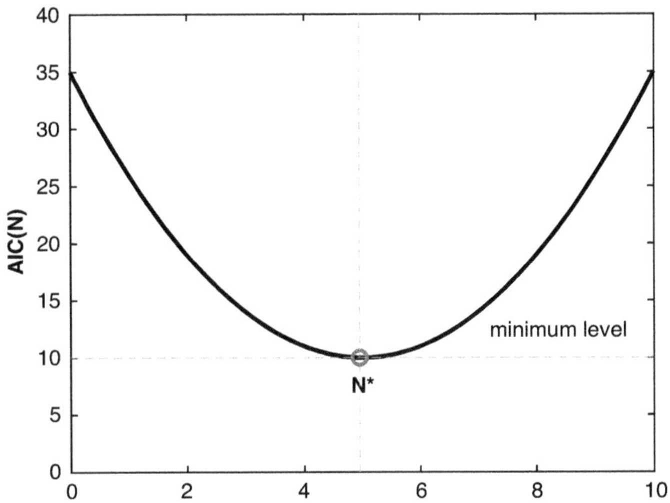

FIGURE B.2 Determination of optimal order by AIC.

In many applications, AIC is complemented by additional checks such as residual whiteness, stability of the selected order across segments, and alternative criteria (e.g., the Final Prediction Error, FPE) to ensure the chosen model is both accurate and robust (Figure B.2).

Bibliography

Akaike, H. (1974). A new look at the statistical model identification. *IEEE Transactions on Automatic Control*, 19(6), 716–723.

Akhyar, S., & Omatu, S. (1993, October). Self-tuning PID control by neural-networks. *Proceedings of 1993 International Conference on Neural Networks (IJCNN-93-Nagoya, Japan)*, vol. 3, pp. 2749–2752. IEEE.

Bengio, Y., Simard, P., & Frasconi, P. (1994). Learning long-term dependencies with gradient descent is difficult. *IEEE Transactions on Neural Networks*, 5(2), 157–166.

Bishop, C. M. (1995). *Neural networks for pattern recognition*. Oxford University Press.

Elman, J. L. (1990). Finding structure in time. *Cognitive Science*, 14(2), 179–211.

Fukushima, K. (1980). Neocognitron: A self-organizing neural network model for a mechanism of pattern recognition unaffected by shift in position. *Biological Cybernetics*, 36(4), 193–202.

Glorot, X., & Bengio, Y. (2010). Understanding the difficulty of training deep feedforward neural networks. *Proceedings of the Thirteenth International Conference on Artificial Intelligence and Statistics (AISTATS 2010, Chia Laguna Resort, Sardinia, Italy)*, pp. 249–256.

Goodfellow, I. J., Shlens, J., & Szegedy, C. (2014). Explaining and harnessing adversarial examples. arXiv preprint arXiv:1412.6572.

Goodfellow, I., Bengio, Y., & Courville, A. (2016). *Deep learning*. MIT Press.

Graves, A. (2012). *Supervised sequence labelling with recurrent neural networks* (vol. 385). Springer Science & Business Media.

Graves, A., Mohamed, A.-R., & Hinton, G. (2013). Speech recognition with deep recurrent neural networks. *2013 IEEE International Conference on Acoustics, Speech and Signal Processing*, Vancouver, BC, Canada, pp. 6645–6649. IEEE.

Haykin, S. (1994). In *Neural networks: A comprehensive foundation*, edited by Harry Wechsler. Macmillan.

He, K., Zhang, X., Ren, S., & Sun, J. (2016). Deep residual learning for image recognition. *Proceedings of the IEEE Conference on Computer Vision and Pattern Recognition (CVPR)*, Las Vegas, NV, USA, pp. 770–778.

Hecht-Nielsen, R. (1992). Theory of the backpropagation neural network. In *Neural networks for perception* (pp. 65–93). Academic Press.

Hines, J. W., Tsoukalas, L. H., & Uhrig, R. E. (1997). *MATLAB supplement to fuzzy and neural approaches in engineering*. John Wiley & Sons, Inc.

Hochreiter, S., & Schmidhuber, J. (1997). Long short-term memory. *Neural Computation*, 9(8), 1735–1780.

Hopfield, J. J. (1982). Neural networks and physical systems with emergent collective computational abilities. *Proceedings of the National Academy of Sciences*, 79(8), 2554–2558.

Hornik, K., Stinchcombe, M., & White, H. (1989). Multilayer feedforward networks are universal approximators. *Neural Networks*, 2(5), 359–366.

https://facultyweb.kennesaw.edu/jdirnber/courses.php.

https://news.mit.edu/2012/explained-sigma-0209.

Ioffe, S., & Szegedy, C. (2015). Batch normalization: Accelerating deep network training by reducing internal covariate shift. *Proceedings of the 32nd International Conference on Machine Learning*, Lille, France. pp. 448–456.

Isman Okieh, O., Seker, S., Akinci, T. C., & Ibrahim Idriss, A. (2024). Optimization of neuro-controller application for maximum power point tracking photovoltaic systems through Shannon's information criteria. *Electric Power Components and Systems*, 1–12. https://doi.org/10.1080/15325008.2024.2328799.

Karpathy, A., & Fei-Fei, L. (2015). Deep visual-semantic alignments for generating image descriptions. *Proceedings of the IEEE Conference on Computer Vision and Pattern Recognition CVPR)*, Boston, MA, USA, pp. 3128–3137.

Kingma, D. P., & Ba, J. (2014). Adam: A method for stochastic optimization. arXiv preprint arXiv:1412.6980.

Kingma, D. P., & Welling, M. (2013). Auto-encoding variational Bayes. arXiv preprint arXiv:1312.6114.

Krizhevsky, A., Sutskever, I., & Hinton, G. E. (2012). ImageNet classification with deep convolutional neural networks. *Advances in Neural Information Processing Systems*, 25, 1097–1105.

LeCun, Y., Bengio, Y., & Hinton, G. (2015). Deep learning. *Nature*, 521(7553), 436–444.

LeCun, Y., Bottou, L., Bengio, Y., & Haffner, P. (1998). Gradient-based learning applied to document recognition. *Proceedings of the IEEE*, 86(11), 2278–2324.

Liu, W., Wang, Z., Liu, X., Zeng, N., Liu, Y., & Alsaadi, F. E. (2017). A survey of deep neural network architectures and their applications. *Neurocomputing*, 234, 11–26.

Mikolov, T., Chen, K., Corrado, G., & Dean, J. (2013). Efficient estimation of word representations in vector space. arXiv preprint arXiv:1301.3781.

Nair, V., & Hinton, G. E. (2010). Rectified linear units improve restricted Boltzmann machines. *Proceedings of the 27th International Conference on Machine Learning*, Haifa, pp. 807–814.

Oord, A. V. D., Kalchbrenner, N., & Kavukcuoglu, K. (2016). Pixel recurrent neural networks. *Proceedings of the 33rd International Conference on Machine Learning*, New York, New York, USA, pp. 1747–1756.

Rabunal, J. R., & Dorado, J. (Eds.). (2006). *Artificial neural networks in real-life applications*. IGI Global.

Rumelhart, D. E., Hinton, G. E., & Williams, R. J. (1986). Learning representations by back-propagating errors. *Nature*, 323(6088), 533–536.

Schmidhuber, J. (1992). Learning complex, extended sequences using the principle of history compression. *Neural Computation*, 4(2), 234–242.

Schmidhuber, J. (2015). Deep learning in neural networks: An overview. *Neural Networks*, 61, 85–117.

Schuster, M., & Paliwal, K. K. (1997). Bidirectional recurrent neural networks. *IEEE Transactions on Signal Processing*, 45(11), 2673–2681.

Seker, S. (1993). An analytical approach based on information theory for neural network architecture. *Proceedings of 1993 International Conference on Neural Networks (IJCNN-93-Nagoya,* Japan), Nagoya, Japan, vol. 1, pp. 309–312. https://doi.org/10.1109/IJCNN.1993.713919.

Seker, S., Akinci, T. C., & Taskin, S. (2012). Spectral and statistical analysis for ferroresonance phenomenon in electric power systems. *Electrical Engineering*, 94, 117–124. https://doi.org/10.1007/s00202-011-0224-4.

Şeker, S., Ayaz, E., & Türkcan, E. (2003). Elman's recurrent neural network applications to condition monitoring in nuclear power plant and rotating machinery. *Engineering Applications of Artificial Intelligence*, 16(7–8), 647–656.

Shannon, C. E. (1948). A mathematical theory of communication. *The Bell System Technical Journal*, 27(3), 379–423.

Silver, D., Huang, A., Maddison, C. J., Guez, A., Sifre, L., Van Den Driessche, G., … Hassabis, D. (2016). Mastering the game of Go with deep neural networks and tree search. *Nature*, 529(7587), 484–489.

Srivastava, N., Hinton, G., Krizhevsky, A., Sutskever, I., & Salakhutdinov, R. (2014). Dropout: A simple way to prevent neural networks from overfitting. *Journal of Machine Learning Research*, 15(1), 1929–1958.

Stufflebeam, R. (n.d.). Neurons, Synapses, Action Potentials, and Neurotransmission. Retrieved from https://www.mind.ilstu.edu/curriculum/neurons_intro/neurons_intro.php.

Sutskever, I., Vinyals, O., & Le, Q. V. (2014). Sequence to sequence learning with neural networks. In *Advances in neural information processing systems* (pp. 3104–3112). https://arxiv.org/abs/1409.3215

Szegedy, C., Liu, W., Jia, Y., Sermanet, P., Reed, S., Anguelov, D., … Rabinovich, A. (2015). Going deeper with convolutions. *Proceedings of the IEEE Conference on Computer Vision and Pattern Recognition (CVPR)*, Boston, MA, USA, pp. 1–9.

Tsoukalas, L. H. (2023). *Fuzzy logic: Applications in artificial intelligence, big data, and machine learning*. McGraw Hill.

Tsoukalas, L. H., & Uhrig, R. E. (1996). *Fuzzy and neural approaches in engineering*. John Wiley & Sons, Inc.

Vaswani, A., Shazeer, N., Parmar, N., Uszkoreit, J., Jones, L., Gomez, A. N., … Polosukhin, I. (2017). Attention is all you need. *Proceedings of the 31st International Conference on Neural Information Processing Systems (NeurIPS 2017)*, Long Beach, California, pp. 5998–6008.

Werbos, P. J. (1974). Beyond regression: New tools for prediction and analysis in the behavioral sciences. Ph.D. dissertation, Harvard University.

Zaremba, W., Sutskever, I., & Vinyals, O. (2014). Recurrent neural network regularization. arXiv preprint arXiv:1409.2329.

Zeiler, M. D., & Fergus, R. (2014). Visualizing and understanding convolutional networks. In *European Conference on Computer Vision* (pp. 818–833). Springer.

Zhang, X., Zhao, J., & LeCun, Y. (2015). Character-level convolutional networks for text classification. In *Advances in neural information processing systems* (pp. 649–657). MIT Press.

Index

Note: **Bold** page numbers refer to tables and *italic* page numbers refer to figures.

V

W

Y

Z

For Product Safety Concerns and Information please contact our EU
representative GPSR@taylorandfrancis.com
Taylor & Francis Verlag GmbH, Kaufingerstraße 24, 80331 München, Germany

www.ingramcontent.com/pod-product-compliance
Ingram Content Group UK Ltd.
Pitfield, Milton Keynes, MK11 3LW, UK
UKHW020157260926
475702UK00047B/667
9781041049111